Service-Learning in Theory and Practice

SERVICE-LEARNING IN THEORY AND PRACTICE

THE FUTURE OF COMMUNITY ENGAGEMENT IN HIGHER EDUCATION

Dan W. Butin

Foreword by Elizabeth L. Hollander

SERVICE-LEARNING IN THEORY AND PRACTICE

First published in 2010 by
PALGRAVE MACMILLAN® in the United States – a division of
St. Martin's Press LLC, 175 Fifth Avenue, New York, NY 10010.

Where this book is distributed in the UK, Europe and the rest of the world, this is by Palgrave Macmillan, a division of Macmillan Publishers Limited, registered in England, company number 785998, of Houndmills, Basingstoke, Hampshire RG21 6XS.

Palgrave Macmillan is the global academic imprint of the above companies and has companies and representatives throughout the world.

Palgrave® and Macmillan® are registered trademarks in the United States, the United Kingdom, Europe and other countries.

ISBN: 978–0–230–62250–0 Hardcover
ISBN: 978–0–230–62251–7 Paperback

Library of Congress Cataloging-in-Publication Data

Butin, Dan W. (Dan Wernaa)
Service-learning in theory and practice: the future of community engagement in higher education / Dan W. Butin; foreword by Elizabeth L. Hollander.
p. cm.
Includes bibliographical references and index.
ISBN 978–0–230–62251–7 (alk. paper)
1. Service learning—United States. 2. Student volunteers in social service—United States. 3. Community and college—United States. 4. Education, Higher—Curricula—United States. 5. Experiential learning—United States. I. Tittle.

LC221.B88 2010
361.3'7—dc22

2009039967

A catalogue record of the book is available from the British Library.

Design by MPS Limited, A Macmillan Company

First edition: March 2010

10 9 8 7 6 5 4 3 2 1

Transferred to Digital Printing in 2012

CONTENTS

List of Figures and Tables

Figures

Tables

Foreword

Elizabeth Hollander

The drive to reassert the civic mission of higher education is now a quarter of a century old. Much has been accomplished in the last twenty-five years both to increase attention to educating the next generation of active citizens and to enhancing the role of colleges and universities as good citizens in their own communities. Membership in Campus Compact (a national coalition of college presidents committed to the civic mission of higher education) has grown from the founding three colleges of 1985 to a 2008 membership of over 1,100 campuses, or a quarter of higher education institutes. In 2006, 91 percent of Campus Compact member campuses offered service learning courses and 85 percent had at least one full-time person devoted to coordinating student service or service-learning opportunities. According to three national surveys by the American Association of Community Colleges, 60 percent of community colleges offer service-learning courses.

Many other organizations and initiatives, such as the American Democracy Project of the American Association of State Colleges and Universities, have been formed to support the civic engagement activities of various sectors of higher education. This effort counts nearly 200 active campuses seeking to educate students in their democratic responsibilities. Campus Community Partnerships for Health is a driving force for civic engagement in all aspects of medical education. Imagining America is a national consortium of eighty colleges and universities committed to public scholarship in the arts, humanities, and design. The land grant colleges now hold an annual self-organized outreach scholarship conference. Research universities, under the leadership of Tufts University and Campus Compact, have formed The Research University Civic Engagement Network (TRUCEN) that meets annually and has issued a two-part report entitled "New Times Demand New Scholarship." In the liberal arts sector, the Pericles Project, founded by the philanthropist Eugene

Lang, serves twenty-two liberal arts colleges interested in teaching students civic engagement. In 2007 the more than two dozen regional and national conferences that were convened on the topic of civic engagement and service-learning had a collective enrollment of more than 18,000.

There has even been significant progress in institutionalizing civic engagement through accreditation, classification, and ranking schemes. In 2006 the Carnegie Foundation for the Advancement of Teaching (CFAT) initiated a voluntary classification for "institutions of community engagement." Of the eighty-eight campuses that applied, seventy-six received this designation. In 2008 an even larger number applied. Several major accreditation boards, including the Western Association of Schools and Colleges (WASC) and the North Central Association (NCA), have civic engagement criteria. *US News and World Report* cites colleges for active learning strategies, including service learning. Even a new controversial *Washington Monthly* ranking of colleges is evidence of the ascendancy of the idea of the civic purposes of higher education. It ranks colleges on both measures of social mobility and student public service. While many would question how meaningful the measures are, attention to them is itself telling.

Another sign of the success of the civic engagement movement is that a number of campuses have obtained significant endowments for their service-learning and civic engagement activities. Duke University received a gift in 2008 of $15 million from the Bill and Melinda Gates Foundation that was matched with $15 million from the Duke Endowment for Charlotte to start the Duke Center for Civic Engagement. Amherst College started a Center for Community Engagement in 2007 with a $13 million endowment gift from the Argosy Foundation (founded by the Boston Scientific chairman, John E. Abele). Tufts University received a $40 million dollar endowment for its College of Citizenship and Public Service from Jonathan Tisch in 2006. In 2005 Harrison Steans endowed the Irwin Steans Center for Service Learning at DePaul University with $5 million and in 2001 Eugene Lang gave Swarthmore $10 million to support the Lang Center for Civic and Social Responsibility.

Yet, even with all of these signs of progress, there are those who are asking if the civic engagement movement in higher education has "stagnated and dissipated" particularly in its aim of transforming institutions to become generators of "democratic, community-based knowledge and action" (Saltmarsh and Hartley 2008, 1). In 2004, the Wingspread conference questioned whether America's colleges and universities were truly ready to commit to engagement and the "radical, institutional change such a commitment will require" (Brukardt et al. 2004, 17). Among the list of radical changes needed were breaking down the barriers of departmental governance, financial incentives for engagement, community partnerships with shared governance, and formal and informal recognition systems within and across higher education for engaged teaching and scholarship. Amy Driscoll wrote in an article in *Change* magazine in 2008 that institutions applying for engagement designation from CFAT were struggling with establishing genuine reciprocity with their communities. She also noted that few institutions had changed the recognition and reward system for promotion and tenure. "In contrast, most institutions continue to place community engagement and its scholarship in the traditional category of service and require other forms of scholarship for promotion and tenure" (Driscoll 2008, 41).

Dan Butin's book is very valuable as we evaluate how far the civic engagement "movement" in higher education has come and what strategies should be considered next. He lays bare the difficulties in relying on service-learning as the standard bearer for a revolutionary redefinition of the nature of scholarship and institutional transformation. He does so from the perspective of a passionate belief in the beneficial impact of service-learning that is truly "antifoundational" and profoundly disruptive of how students normally acquire knowledge and from whom. Antifoundational service-learning requires students, as Butin says later in this book, to ask fundamental questions about justice, to hear voices rarely heard (e.g., prisoners) and reveal the "'deep divisions' within which and through which we think about content knowledge, cultural openness, and oppression." Butin challenges the technical, cultural, and political limits of most types of service-learning to accomplish this kind of antifoundational learning, facing up to all the hard questions that have been posed over the last decade. Does service-learning truly privilege experiential learning over disciplinary content? Does service-learning effectively develop multicultural competence? Is service-learning dominated by left liberal politics? Does service-learning fulfill its promise of decentering static and singular notions of teaching, learning, and research by moving against the grain of traditional practice? Or, instead, is it likely to be dissipated by being "misappropriated and drained of its transformative power"?

Butin suggests an incremental strategy for changing higher education that calls for relying on a disciplined approach to thinking deeply about the theory and practice of service-learning. One way to achieve this, he suggests, might be to embed service-learning in a disciplinary home such as community studies.

Butin is a singular voice in the movement. He combines a radical view of the most profound manifestations of service-learning and an incremental view of service-learning as the political lever to transform higher education. He does not see disciplined thought as a threat to service-learning. Instead he urges bringing the academic life of examining, questioning, researching, and synthesizing to bear on making service-learning better, deeper, and, in the end, more effective and more widely understood by faculty. He believes that this approach will make it more likely that faculty will "embed community-based models of teaching, learning, and research into the very structure and thus practices of our jobs" in the long run. Butin uses the history of women's studies as an example of transformation that was accomplished by creating a disciplinary space for deep discussion that, over time, affected how the entire academy treated women's voices.

As long as the service-learning movement has been around, there has been debate about how, or even whether, change should be moved from the "margins to the mainstream." The "Invisible College," now known as Educators for Community Engagement, was formed in 1992 to "provide a free space for faculty to explore the difficult issues raised by service learning—issues of pedagogy and of responsible relationships with communities" (http://www.e4ce.org/About/History.htm, accessed on June 2, 2008). The group originally called itself The Invisible College as it made up an invisible network of faculty in various colleges who were committed to service-learning. Fostered by both Campus Compact and the Campus Outreach Opportunity League (COOL), this pioneering faculty group sought to create forums that included the voices of community members and students to think deeply about what was, and remains, a challenging pedagogical approach. They rejected the traditional conference format of papers, presentations, and commentary, and instead organized "national gatherings" structured in the form of learning circles.

By the time I came to Campus Compact in 1997, there was controversy about whether The Invisible College would be an effective strategy for spreading the practice of service-learning. Some leaders of the movement were concerned that these faculty would be seen as a radical "fringe" group that would "scare off" more mainstream faculty who were being introduced to service-learning.

There are those who will be similarly worried that Butin's suggestions to "discipline" service-learning will, in fact, marginalize the practice. It will be argued that to place service-learning in a discipline such as community studies will signal to faculty in other disciplines that the practice belongs only in that department, rather than allow viewing it as a pedagogical approach that can be used in any disciplinary subject. On the other hand, everyone can agree that the impact of service-learning on both students and the communities served depends greatly on its quality. We also know that the current quality of service-learning teaching varies a lot, so much so that people speak of a norm for community collaboration of "do no harm." There is danger in providing students with a "drive by" community experience that does not address issues of power and privilege.

Improving quality in any practice requires a "safe space" to examine the aims of the practice, its potential to achieve those aims, and its limitations within the current context of higher education. It is hard to find this space when devotees of service-learning are in the midst of a campaign to overturn, or at least modify, long-standing structures that privilege traditional teaching and research practice. Any campaign, however, is improved by reflection on its aims and effectiveness. This book provides a series of hard questions that can frame such reflection. It reminds us of the serious barriers to mounting a revolutionary approach to changing higher education, as opposed to an incremental approach. It also reminds us of the contemporary challenges of higher education, as college becomes more and more of an economic necessity and higher education gets transformed by efforts to make it more "efficient" and cost-effective. Will there be any room for, or interest in, pedagogical strategies designed to "shake up" students and help them understand the extent to which our notions of justice and privilege are socially constructed and can be changed?

Given the seismic changes in higher education resulting from its explosive growth, including the growth of for-profit and virtual campuses, Butin argues that this is a particularly important time to think through the role and impact of service-learning.

Should service-learning carry all the weight of reasserting the civic mission of higher education? Is it realizing its potential to be truly "justice seeking," truly decentering the privileged knowledge in the academy? I would also ask, are we taking advantage of several decades of postmodern theorizing about the nature of knowledge and exploring how service-learning can be informed and deepened by this thinking?

As I look across the landscape, campuses are finding increasingly diverse ways to be engaged and to engage their students. Among these efforts are community-based research, public scholarship, rediscovery of extension services, campus collaborations for urban revitalization, civic leadership programs, social entrepreneurship programs, fostering democratic deliberations, teaching political engagement, and interdisciplinary centers focused on pressing social problems such as global warming, school improvement, and family and community development. This proliferation of efforts can be viewed as a sign of the success of the movement to reassert the civic purposes of higher education. They are, however, being undertaken by all but a very few campuses, within the context of the traditional German model of discipline-based education, and are not, indeed, transforming higher education. Embedding engaged teaching and learning in faculty promotion and tenure systems is slow at best. Does this mean that these efforts may, in some ways, reflect the co-optation and even "misappropriation" of the aims of service-learning that Butin fears?

This is an important time to reassess the vision of and progress toward the engaged campus. Butin's book can help all of us who care passionately about the civic role of the university to think more deeply about it and the role service-learning can play to achieve it if we have the fortitude to hear a "different voice" calling for disciplined thought.

Preface

There is no one thing called service-learning. From this seemingly obvious point, one can begin to tug at the very foundations of the service-learning movement in higher education. In so doing, it becomes possible to see how traditional theories and practices of service-learning have substantial limitations to their efficacy and sustainability; it also becomes possible, though, to rethink and rework the potential of the transformative nature of community-based models of teaching, learning, and research.

This book is an attempt to do exactly this kind of work. It is a sustained inquiry into what it might mean that there is no one thing called service-learning. For if we acknowledge the multiple, divergent, and often contradictory modes by which we do service-learning in higher education, what we can truly say is that there is no singular, unitary, or obvious model by which service-learning is or should be done. This move—of undercutting the surety of our practices —lays bare that the theory of service-learning is actually a set of theories contingent on the embodied and experiential character of the service-learning experience.

I spend this book examining and explaining what this means and the implications thereof. But to be brief, let me suggest that it ultimately means that the service-learning movement is in need of a new generation of scholarship that carefully and critically examines the gaps, limits, and problematics of an incredibly complex practice with no singular core metanarrative. My argument is that the service-learning movement has been positioned as an overarching answer for higher education, and, in so doing, has found itself within an ever-tightening double bind. In its quest to be singular, universal, and obvious—what I term the service-learning-as-social-movement phenomenon—the service-learning movement has all too often become ineffectual in accomplishing its intended transformational goals. Yet any attempts to reclaim the power of community engagement raises the specter of critique that the movement cannot account for. I thus suggest that we must think through what an "academic home" for service-learning might look like and accomplish, what I term service-learning-as-intellectual-movement.

This is not the way that we traditionally experience or read about or research the theory or practice of service-learning. Rather, service-learning—the linkage of academic work with community-based engagement within a framework of respect, reciprocity, relevance, and reflection—has become an extremely popular form of active pedagogy and civic(s) education in higher education. It is understood as part of the larger community engagement movement in higher education—a set of loosely interrelated practices and philosophies such as civic engagement, community-based research, public scholarship, and participatory action research—that is traditionally viewed as the successful linkage of classrooms with communities and theory with practice that improves students' academic achievement, enhances their cultural competence, and fosters a more inclusive and just world.

Service-learning, in particular, has become a standard-bearer for this type of engaged learning that embraces the possibilities of conjoint civic renewal and academic betterment. This is, in fact, a wonderful affirmation of the development of service-learning in higher education over the last thirty years. What began as a set of ad hoc practices in the civil rights era (Stanton et al. 1999) has transformed into a set of institutionalized practices found across the majority of colleges and universities. It is seen as both a pedagogy and philosophy that links classrooms with communities and textbooks to the "real world." At its best, service-learning is seen as an embodiment of what Ernest Boyer (1990, 1996) termed a "scholarship of engagement" that tightly links research, teaching, and activism to foster individual, institutional, and community change (Colbeck and Michael 2006; Harkavy 2006).

Recent reports (e.g., HERI 2009; Lindholm et al. 2005; NSSE 2007), in fact, show that a vast majority of faculty believe that creating and sustaining partnerships with local communities is of "high" or "highest" priority for their institution and that the practice of service-learning is a prominent marker and predictor of students' "deep learning." Moreover, the Carnegie Foundation for the Advancement of Teaching has—in an acknowledgement of and attempt to systematize such developments—created an elective classification of "community engagement" that allows institutions to document how their teaching and learning models and research practices and priorities align with and support collaborative community practices (Carnegie 2006; Driscoll 2008).

Service-learning thus appears well positioned to flourish in higher education. More than 1,100 postsecondary institutions are members of Campus Compact, a national organization of college and university presidents committed to community service, civic engagement, and service-learning in higher education. A preponderance of research suggests that service-learning is a positive and statistically significant pedagogical intervention across a wide range of outcome variables, such as students' academic outcomes, openness to diversity, and postcollege civic engagement (Eyler et al. 2001; Bell et al. 2007). And the practice has become a fashionable and coveted marker in higher education that draws prospective students, inspires donors, fosters exciting pedagogical opportunities, and positions institutions for federal grant dollars.

Such positive developments can be clearly seen within the service-learning field as scholarship and debate has begun to move from issues of legitimation to those of institutionalization. Rather than attempting to debate its merits, scholars now focus on the modes and means by which service-learning can be embedded across academic disciplines and within instructional norms and institutional policies (e.g., Furco 2002; Hartley et al. 2005). Put otherwise, Boyer's call for a "scholarship of engagement" has flowered into the implementation of a potpourri of complementary and overlapping pedagogical practices, as a wide range of scholars in higher education have called for a more public and participatory scholarship, one that is academically rigorous as well as civically engaged, useful, and responsive (Colby et al. 2003, 2007; Ramaley 2006; Woodrow Wilson Foundation 2005).

Yet this book suggests that the theory and practice of community engagement in higher education is reaching an upper limit, an "engagement ceiling" that ultimately limits the potential power of the engaged campus.

This book thus rethinks how we talk about and engage with our local and global communities, and, in turn, puts forward alternative practices and theories for moving forward. Doing so may cause quite some consternation, for it challenges our by-now taken-for-granted presumptions of a pedagogy and philosophy (i.e., service-learning) that itself began in the 1960s and '70s as an attempt to undermine the taken-for-granted assumptions of an academy seemingly out of touch with its sense of purpose and unreflective about its own pedagogical practices. Yet this second wave of critique, I suggest, is crucial if the community engagement movement is to ultimately flourish in higher education.

This is, I suggest, a critical moment in the community engagement movement's attempt toward legitimacy and institutionalization. Massive demographic shifts, marketization and globalization, and a fundamental rethinking of the values and mission of higher education have caused both anxiety and renewed fervor for the articulation of higher education for the "public good" (Kezar et al 2005; see also Benson et al. 2007; Colby et al. 2007; Kirp 2003; Zemsky et al. 2005). As the notion of a "traditional" college education continues to fracture even as it becomes an ever clearer pathway for future success, stakeholders question and reaffirm what it means to imbue students with a sense of inquiry and critical thinking, civic engagement and social responsibility, and integrative and interdisciplinary perspectives of the world (AAC&U 2007).

It is against such a backdrop that this book is written. Specifically, this book is a careful and critical analysis of the limits and possibilities of the theory and practice of service-learning in higher education. It is a critical investigation into what may be possible, and what may not be, in institutionalizing and implementing service-learning across the higher education landscape. My goal is not one of disparaging or detracting from the service-learning movement. It is rather an attempt to strengthen the long-term viability of this potentially transformative mode of engaged pedagogy by rethinking and expanding a vision of community engagement within higher education.

Specifically, I suggest that the rapid expansion of service-learning in higher education (particularly in the last decade) has overtaken the field's ability to examine and account for the implication of its success. The strategy of service-learning-as-social-movement that sustained and nurtured its early years no longer suffices to inform contemporary complexities and problems (see, e.g., Head 2007; Hogan 2002; Schutz 2006; Swaminathan 2007). I lay out the specific arguments and data throughout the book in order to suggest that there are very particular core fissures that have specific conceptual, institutional, political, and pedagogical roots. I explore these in-depth to trace their origins and to shift the tactical and theoretical foundations of how we think about and engage in service-learning and community engagement. A scholarly criticality is thus necessary to engage with these issues. There are three distinct reasons for the need to be critical of service-learning.

First, if we do not understand the limits of service-learning in higher education, we cannot understand its possibilities. Our practices and strategies in supporting service-learning are guided by the assumptions we carry of teaching, learning, and scholarship. The desire to develop more engaged and transformative teaching models is a noble and necessary one; yet it often ignores or glosses over a wide range of educational research that has documented both teacher and student resistance to such practices. It minimizes the risk (pedagogical, institutional, and existential) of engaging in a paradigmatically different form of teaching and learning, and it undertheorizes the implications of sustainability of such a transformative model. To ignore these blind spots is to remain beholden to them.

Second, without critique, the service-learning movement will remain beholden to a destructive and debilitating binary thinking concerning its own advancement. John Dewey (1938) famously began his *Experience and Education* with the observation, "Mankind likes to think in terms of extreme opposites. It is given to formulating its beliefs in terms of *Either-Ors*, between which it recognizes no intermediate possibilities" (17). This stance in the service-learning movement arises from the desire to sustain the growth and institutionalization seen in the last decade; critique, from this perspective, undercuts a genuine transformational movement. Yet the maintenance of the dichotomy of being "for" or "against" service-learning oversimplifies an immensely complex pedagogical practice and undermines its own affinity to the academy, where criticality is the norm. The very definition of scholarship is that it is public, able to be critiqued, and able to be built upon (see, e.g., Shulman 2004). Critique thus allows nuance, subtlety, and alternative directions to emerge.

Third, if the service-learning movement does not itself take up the task of investigating, critiquing, and resolving its own issues, then others, with less noble motivations, will do so. A prominent example of this is the ongoing debate of "intellectual diversity" in the academy. For example, college students, conservative organizations argue, are only "getting half the story" since faculty are predominantly liberal and the courses offered tend toward "indoctrination rather than education" (ACTA 2006). While service-learning is not inherently "political," much of its practice and many of its proponents implicitly and explicitly link it to causes of social justice. It is thus incumbent on the field to explore and debate how it can support its own practices and policies both internally and to the larger public.

Critique, of course, is never easily taken. For critique implies that there is something wrong about the service-learning field attempting to make a difference. But it is exactly critique, this book argues, that will sustain service-learning. The very fact that resistance to critique is present in the service-learning field signifies the very fragility of the foundations for service-learning in higher education. Service-learning cannot sustain itself in higher education—which is built on the very basis of rational and sustained critique and examination of inconvenient truths—if it cannot accommodate itself to the functioning of the academy.

Put positively, critique is exactly what can support service-learning. For it offers an opportunity to reexamine and rethink service-learning as an immensely powerful form of pedagogy for undercutting our sense of the normal and taken-for-granted perspective of the world and ourselves. It is this goal that the book is after. For service-learning and other forms of community engagement offer a means to engage in powerful pedagogical and research practices that foster questioning and doubt and that can lead to students' rethinking of themselves and their view of the world (Baldwin et al. 2007; Borden 2007; Paoletti 2007).

This book is thus an attempt to clear the ground for a rethinking and recommitment to transformative notions of service-learning and community engagement. I do so through three specific steps that are mirrored in the three-part structure of this book: examining the internal and external limits and challenges to service-learning; articulating complementary and alternative models of community-engaged teaching, learning, and research within a disciplinary context; and rethinking a means for strengthened community engagement in higher education.

The first part of this book maps out the limits and possibilities of service-learning in higher education. It examines the multiple barriers that limit the institutionalization of service-learning, and provides a theoretical and empirical foundation for its potential as a transformational pedagogy. I do so by first clarifying what we mean when we talk about service-learning; for in fact there are four distinctive ways in which service-learning is "used"—technical, cultural, political, and antifoundational. Such clarification makes it possible to see the limits and possibilities of each distinctive mode, as well as to understand how these conceptions overlap and oftentimes implicitly conflict with each other. Put otherwise, most models of service-learning undercut and subvert themselves because of their very own internal assumptions and functioning. I conclude this first section of the book by suggesting how we may begin to move forward given the articulated constraints by viewing service-learning as a disruptive practice that can in fact be a form of justice-oriented education that I term "justice-learning"—service-learning as a self-reflexive pedagogy that engages rather than closes off (and thus reifies) the very categories it is meant to operate within.

In the second part of the book, I develop an alternative framework within which to think about service-learning. Namely, I suggest that the institutionalization of service-learning may be most successful within a disciplinary "academic home." I explore this argument through a theoretical articulation; an empirical examination of programs that already have certificates, minors, and majors in service-learning; and by examples and case studies of programs (within and outside of the service-learning field) that have attempted to institutionalize and legitimate themselves through disciplinary means. Such realizations allow not only a reframing of some of the grounding assumptions of service-learning, but a means for articulating genuine and powerful programs that link academic coursework with long-term, immersing, and meaningful community engagement for sustainability and transformation within higher education.

The concluding part of the book examines the larger vision of community engagement in higher education. It details the means for strengthening of faculty buying into community engagement, offers a theoretical and empirical grounding, and outlines the growing challenges for sustained and consequential community engagement as higher education grapples with new realities such as the increasing racial diversity of students, external market pressures, and the changing terms of faculty employment.

Near the end of his life, the French philosopher Michel Foucault (2000) argued in an interview that, "I'm an experimenter in the sense that I write in order to change myself and in order not to think the same thing as before" (240). In the same vein, this book is an attempt to experiment with rethinking the possibilities of service-learning in higher education. For the service-learning movement has attempted to position itself exactly as a theoretically and pedagogically unproblematic practice to be embedded within higher education. However, the center will not hold. For the academy is by its very nature a space for examination and critique, especially when confronted with issues as complex and contested as what transpires within and across communities. It is thus incumbent on the service-learning field to carefully and critically examine its own practices and theories to strengthen them rather than have them picked apart by not-so-gentle critics.

So if we are to begin to think carefully, critically, and differently about service-learning, if we are "not to think the same thing as before," then I would argue (with Foucault) that we must experiment. We must experiment with what service-learning could be. This book engages issues that are, I hope, provocative, critical, and disruptive examinations of service-learning. I promote these to avoid complacency within a field that has been blessed (and thus perhaps cursed) with a decade-long expansion into an academy of which it is yet not truly a part. This book offers an opportunity to experiment with rethinking and constructing a service-learning made stronger by critique.

This book represents a synthesis of my work over the last five years on issues of service-learning and community engagement. It is an attempt to bring together some of my past work (much of which has been extensively modified and updated) with my more recent examinations and thinking. There has been, I believe, a sustained and overarching argument throughout my writings over the years; but this is the first time that I have been able to put them all together into what I consider to be a comprehensive and sustained argument.

Thanks for this are many. First to Julia Cohen, my editor at Palgrave, who helped me to move this book to publication in a timely manner and believed in the need for such a contribution to the field. Second, to a small but supportive group of colleagues within the service-learning field who encouraged me to continue such writings and questionings. Jeff Howard, Kelley Skillin, Amy Smitter, Jen Gilbride-Brown, and Katharine Kravetz provided encouragement at key stages. Katharine, in particular, was a wonderful resource and interlocutor as I struggled to articulate my critiques and questioned, along the way, whether such criticisms were valued by my readers.

And it is here that I must especially thank Liz Hollander. As a former president of Campus Compact, Liz was at first skeptical of my work and, by extension, of my commitment to service-learning and community engagement. I had just published my article "The Limits of Service-Learning in Higher Education." This was more than a good enough reason, it would seem, to be skeptical. So after a few emails we finally met at an Au Bon Pain in Harvard Square. We talked for two hours and realized that we were both after the same thing: a strong and sustained future for service-learning in higher education. Since that initial meeting we have had wonderful conversations in person, by phone, and by email. Liz has become one of my most careful, sympathetic, and critical readers. She is of course quick to tell me when she disagrees with my assumptions and conclusions; but she has never backed away from the debate. And, I believe it is fair to say, we have both changed our perspectives accordingly along the way. It speaks volumes about Liz that she has had the openness to engage in this discussion, and I am thankful to her for that.

Finally, I thank my wife, Gitte Wernaa Butin, for being the best and most critical reader of who I am and thus of my work. The finishing of this book is dedicated to her and thus to us.

Part I

Defining and Disturbing Service-Learning in Higher Education

Chapter 1

Conceptualizing Service-Learning

What exactly is service-learning? A pedagogy? A philosophy? And what are its possibilities? Can it change an individual's perspective? Can it transform a classroom? A community? And, perhaps most critically, what are its limits? The first part of this book examines these questions in-depth, with this chapter focusing on carefully and analytically conceptualizing what service-learning is (and is not), laying the groundwork for the analyses in the rest of the book, and drawing out some implications for the field of service-learning in general.

The standard argument is that service-learning pedagogy rejects the "banking" model of education where the downward transference of information from knowledgeable teachers to passive students is conducted in fifty-minute increments. It subverts the notion of classroom as graveyard—rows and rows of silent bodies—for an active pedagogy committed to connecting theory and practice, schools and community, the cognitive and the ethical. Such an active and engaging framework has garnered national attention as a means of reengaging today's students with both academics and civic values (Colby et al. 2003, 2007).

Moreover, service-learning advocates point to research demonstrating that service-learning enhances student outcomes (cognitive, affective, and ethical), fosters a more active citizenry, promotes a "scholarship of engagement" among teachers and institutions, supports a more equitable society, and reconnects colleges and universities with their local and regional communities (e.g., Astin et al. 1999; Benson et al. 2007; Wade 2007). By emphasizing real-world learning and reciprocity between postsecondary institutions and communities, service-learning seems to serve as a powerful counterpoint to contemporary positivistic educational trends that deprofessionalize teaching, narrowly focus on quantifiable outcomes, and maintain instrumental conceptions of teaching and learning.

Yet despite (or perhaps because of) the recent proliferation and expansion of service-learning theory and practice, there is a troubling ambiguity concerning even basic principles and goals in the service-learning literature. Is service-learning a pedagogical strategy for better comprehension of course content? A philosophical stance committed to the betterment of the local and/or global community? An institutionalized mechanism fostering students' growth and self-awareness concerning issues of diversity, volunteerism, and civic responsibility? Or, as some critics note, a voyeuristic exploitation of the "cultural other" that masquerades as academically sanctioned "servant leadership" (e.g., Cross 2005; McKnight 1989)? All of these perspectives, and more, are to be found as prominent articulations of what service-learning "truly" is.

I thus want to begin to clarify service-learning practice and theory by offering four distinct conceptualizations of how service-learning is articulated in the literature and enacted in the field: technical, cultural, political, and antifoundational. In doing so, I hope to accomplish three goals: first, to clarify the assumptions of and implications for service-learning within each perspective; second, to suggest that the dissonance and synthesis across these multiple perspectives offers a means of understanding some of the most vexing problems within service-learning theory and practice; and third, to demonstrate that service-learning is never a singular, stable, or, ultimately, controllable practice. This last point, in particular, underpins this book's approach to rethinking our approach of how we think about, talk about, and enact service-learning and other forms of engaged scholarship in the academy.

CONCEPTUALIZING SERVICE-LEARNING FROM MULTIPLE PERSPECTIVES

Scholars have put forward useful definitions, criteria, and conceptualizations of service-learning. A commonly cited definition (Bringle and Hatcher 1995) argues that "service-learning [is] a course-based, credit-bearing, educational experience in which students (a) participate in an organized service activity that meets identified community needs and (b) reflect on the service activity in such a way as to gain further understanding of course content, a broader appreciation of the discipline, and an enhanced sense of civic responsibility" (112). Such an articulation is a model in the field precisely because it attempts to balance service and learning and link them in a meaningful way.

It is possible, other scholars suggest (Furco 1996; MJCSL 2001; Sigmon 1994), that a spectrum of service programs fall under the rubric of "service-learning." Programs that emphasize the service component and the "served"—for example, volunteer activities and community service—would fall on one end of the spectrum, while programs that focus on the learning and the "provider" of the services—for example, internships and field-based education—would fall on the other end. The scope of what potentially counts as service-learning has thus resulted in the development of multiple monikers—"academic service-learning," "community-based service-learning," "field-based community service"—in an attempt to differentiate between programs and emphasize what is of primacy.

Irrespective of the definitional emphasis, service-learning advocates put forward a consistent articulation of the criteria for service-learning to be legitimate, ethical, and useful. These may be glossed as the "4 Rs"—respect, reciprocity, relevance, and reflection (Sigmon 1979; MJCSL 2001; Campus Compact 2000). First, those doing the serving should always be respectful of the circumstances, outlooks, and ways of life of those being served. The point to be made is that the server is not a "white knight" riding in to save anyone but just another human being who must respect the situation she is coming into. Second, the service is not to benefit only the server—for example, the white, middle-class, preservice teacher who, through her tutoring, becomes exposed to and begins to understand how the "underprivileged" live and behave. Not only should the server provide a meaningful and relevant service to those he is serving, but often members of the community being served should be the ones responsible for articulating what the service should be in the first place.

Third, the service must be relevant to the academic content of the course. This is not simply to say that course credit is based upon learning rather than service; more forcefully, the service should be a central component of a course and help students engage with, reinforce, extend, and/or question its content. Finally, service-learning does not provide transparent experiences; reflection is required to provide context and meaning. Given the real-time aspect of service-learning, students need multiple opportunities to engage with the ambiguity and complexity of the experience. The issues that arise, for example, from tutoring migrant youth or working with elderly hospice patients are not simple topics that can be addressed in a forty-five- or even a ninety-minute class; they require time for reflection, discussion, and research.

The breadth of how and why service-learning is enacted—it is used from elementary school until graduate school and in disciplines ranging from accounting to women's studies—provides for a wide range of conceptualizations of the field. Kendall (1990), for example, differentiates between service-learning as a pedagogy—a specific methodology for the delivery of content knowledge—and as a philosophy—a world view that permeates the curriculum, instruction, and assessment of a course. Alternatively, Lisman (1998) suggests that all modes of service-learning are embedded in philosophical orientations that he differentiates as "volunteerism," "consumerism," "social transformation," and "participatory democracy." Each perspective, Lisman argues, privileges different modes of service-learning engagement with vastly different impacts on individual and societal outcomes. Other scholars (e.g., Morton 1995; Liu 1995) attempt to avoid such dichotomization by suggesting that all modes of service-learning, if enacted "thickly" enough, are useful for providing valuable service and increasing academic learning.

At one level, I find these definitions, criteria, and conceptualizations helpful. Such perspectives provide a useful heuristic for understanding and contrasting distinct and often divergent forms of service-learning as they are perceived and enacted across multiple disciplines. They provide both insiders (e.g., practitioners, researchers, tenure committees, and administrators) and outsiders (e.g., general public, philanthropic organizations, and lawmakers) a language for situating the multiple forms of service-learning. Yet there are several distinct problems with such traditional articulations of how to understand service-learning: the problematics of a latent teleology and an unsupportable ethical foundationalism.

By a latent teleology I mean that most definitions, criteria, and conceptualizations of service-learning privilege particular modes and goals of service-learning and view deviations from such implicit norms as derivative. Just like the medieval concept of an ordered universe structured around a "great chain of being," service-learning scholarship and practice privileges: (a) volunteer activities done by (b) individual students with high cultural capital for the sake of (c) individuals with low cultural capital (d) within the context of an academic class (e) with ameliorative consequences. To reverse any of these five preconditions or the seemingly logical flow of the direction of these actions is to expose the strong normative framework within which service-learning operates.

To suggest, for example, that students engaged in service-learning be paid, or that they provide service to the rich, or that the outcomes may be other than positive, is to go against the grain of the implicit normative framework of what is understood by "service-learning." My point is not to promote any such specific alternative for the moment; rather, it is to make clear that undergirding almost all conceptualizations of service-learning are modernist, liberal, and radical individualistic notions of self, progress, knowledge, and power.

Specifically, such a worldview is grounded in the notion that individuals are autonomous change agents who can effect positive and sustained transformations. It is the belief that we can consciously and deliberately bring about betterment (by the more powerful for the less powerful) through a downward benevolence whereby all benefit. This is the Enlightenment project writ large, grounded in a latent teleology that presumes and works within a worldview of a constantly progressive upward movement through seemingly universal goodwill and good faith.

The latter point also reveals the unsupportable ethical foundationalism of service-learning. Namely, service-learning practice and scholarship is predicated on the belief that both the process and outcomes of service-learning are universally beneficent. This view seems confirmed when one hears stories of elementary school students revitalizing run-down neighborhoods or college students working hand-in-hand with community organizations to develop environmental impact statements concerning a proposed incinerator plant. It becomes more problematic, however, when certain kinds of questions are made visible: What sustained community impact is achieved? Who benefits from the enactment (and publicity) of such processes? What actual learning is documented as a result of such a process? Service-learning programs, to take but one example, have promoted much goodwill among those doing the actual service-learning, but there is considerably less evidence that it has provided much benefit for the recipients.

It is sufficient at this point to note that both through the frame of critical theory and the "posts" (postmodernism, poststructuralism, and postcolonialism [Kumashiro 2002]), all the above-stated suppositions have become problematic. From the notion of autonomous individuals consciously willing positive change to the win-win mantra of service-learning advocates, conceptualizations of service-learning have glossed over the presumption of neutrality, the privileging of whiteness, and the imbalance of power relations. I will address these issues and limits in much more detail in the next two chapters; for now it is enough to acknowledge these critiques to articulate a different means by which to view service-learning.

It is thus useful to put forward four perspectives on service-learning—technical, cultural, political, and antifoundational—that can differentiate between highly distinctive and divergent goals. I should note that I do not presume that service-learning practice and scholarship neatly separates into four ideal types; rather, I put forward such distinctive perspectives as a means of clarifying what is possible within the service-learning field. Additionally, I acknowledge that such typologies overlap, blend, and are reconstituted in a multiplicity of unanticipated modes. Nevertheless, it is critical that some basic premises and overarching paradigms are delineated to bring greater clarity to service-learning scholarship and practice. Moreover, differentiating such four distinctive conceptualizations of service-learning offers a more robust model for understanding the undergirding assumptions and implications of service-learning. This has, as I will show throughout this and following chapters, numerous implications for how we understand faculty buy-in, student engagement, community involvement, and a host of other critical components of the institutionalization and transformative power of service-learning in higher education.

A Technical Perspective on Service-Learning

A technical perspective on educational reform focuses on "the innovation itself, on its characteristics and component parts and its production and introduction as a technology" (Hargreaves et al. 2002, 73). Questions concerning an innovation's legitimacy and implications are muted or even absent. Rather, technical considerations of implementation are highlighted. Thus questions of efficacy, quality, efficiency, and sustainability of both the process and the outcome of the innovation come to the forefront.

This perspective constitutes a major strand within service-learning scholarship. A host of linkages between service-learning and student outcomes—personal, social, and cognitive—have been analyzed and "best practice" principles have been put forward. Service-learning has been shown to enhance, among other things, students' personal efficacy and moral development, social responsibility and civic engagement, academic learning, transfer of knowledge, and critical thinking skills (Astin and Sax 1998; Astin, Sax, and Avalos 1999; Eyle and Giles 1999; Rhoads and Howard 1998; Markus, Howard and King 1993). Crucial program characteristics of service-learning experiences, irrespective of the academic discipline, include the quality of the placement, the frequency and length of contact hours, the scope and frequency of in-class and out-of-class reflection, the perceived impact of the service, and students' exposure to and interaction with individuals and community groups of diverse backgrounds (Eyler and Giles 1999; Young et al. 2007).

Service-learning scholars have particularly emphasized the need to link service-learning with enhanced cognitive outcomes as the key to legitimize and sustain service-learning in higher education. Over a decade ago, Zlotkowski (1995) argued that the very future of service-learning within higher education may rest upon "a single elusive but nonetheless basic decision—whether the [service-learning] movement as a whole prioritizes ideological or academic issues" (126). The quantitative linkage of service-learning to such qualities as "deep learning" (NSSE 2007) seems to demonstrate the success of a focus on academics. Thus the phrase "academic service learning" (see, e.g., Rhoads and Howard 1998) has become a symbol of the need to demonstrate how service is a means to learning rather than the goal in and of itself. In one review of the research, Eyler (2000) concludes that "we know that service-learning has a small but consistent impact on a number of important outcomes for students. Now we need to push ahead to empirically answer questions about improving the academic effectiveness of service-learning" (16). Such a "tinkering" approach to educational reform—where the goal is primarily to make a known product better through incremental and systematic change—positions service-learning as a legitimate educational practice in need only of more rigorous and sustained research and operationalization.

A Cultural Perspective on Service-Learning

Rather than focusing on the innovation itself, a cultural perspective emphasizes individuals' meaning-making within and through the context of the innovation. Such meaning-making may be broadly understood within what Geertz (1973) famously termed our "webs of meaning" in that we make sense of who we are with respect to both local and global communities. A cultural perspective—at both the micro/individual and macro/societal level—is thus concerned with normative questions of acculturation, understanding, and appropriation of the innovation.

Service-learning is highly amenable to a cultural perspective. From a macro perspective, it can be viewed as a means of repairing what social theorists describe as the frayed social networks of our increasingly individualistic and narcissistic society (Bellah et al. 1986; Putnam 2000). Advocates suggest that service-learning is an ideal means by which to support and extend civic engagement, foster democratic renewal, and enhance individuals' sense of community and belongingness to something greater than themselves (Barber 1992; Lisman 1998). From a micro perspective, service-learning can be seen as a means of fostering in the individual a respect for and increased tolerance of diversity, gaining a greater awareness of societal concerns, developing a stronger moral and ethical sense, and encouraging volunteerism and civic engagement (Coles 1993).

These two levels are linked to the extent that we come to know about ourselves by engaging with those who are different from us. Such greater knowledge, in turn, affects how we think about and engage with the world we live in. It is therefore critical to acknowledge that, from a cultural perspective of service-learning, diversity in the placement site acts as a crucial mediator between individual self-knowledge and societal responsibility (Bell et al. 2007). By engaging with those different from themselves—with "difference" primarily understood across racial, ethnic, class, and sexual orientation lines—students will come to better understand, respect, and engage with the cultural plurality of our diverse society (Henry 2005).

This perspective constitutes a second major strand within the service-learning movement and is often linked to the technical perspective. Thus Eyler and Giles (1999) do not hesitate to include citizenship alongside critical thinking as a legitimate student outcome of service-learning. Likewise, the *Service-Learning Course Design Workbook* put out by the Michigan Journal of Community Service Learning (2001) argues that academic service-learning is, by definition, the linkage of meaningful service to academic and civic learning. Such interweaving of the technical and cultural perspectives is in fact ubiquitous in service-learning literature.

Service-learning is thus understood from both perspectives as a particular methodology for accomplishing specific goals. It should be noted that a cultural perspective mitigates somewhat such an instrumentalist conceptualization; a cultural perspective acknowledges that service-learning outcomes are often embedded within the process itself. As such, a cultural perspective privileges the affective, ethical, and formative aspects of service-learning, and is concerned with linking these experiential components to local, national, and international issues.

A Political Perspective on Service-Learning

A political perspective on an innovation is most concerned with issues of competing constituencies and how these issues are manifest through power (im)balances, questions of legitimacy, allowed and/or silenced perspectives, and negotiations over neutrality/objectivity. It is within a political perspective that an innovation is examined and challenged on normative, ethical, epistemological, and ontological grounds. Whose voices are heard and whose are silenced? Who makes the decisions and by what criteria? Who benefits from such decisions and who loses? To what extent is the innovation a repetition, a reinforcement, or a revocation of the status quo? A political perspective presumes that conflict rather than consensus is an underlying aspect and consequence of the process and product of an innovation such as service-learning.

From a political perspective, service-learning is both potentially transformative and repressive. It is transformative to the extent that education becomes a disruption of the hierarchy and authority of the student-teacher relationship (Freire 1994; hooks 1994). Rather than a didactic "banking model" of knowledge dissemination and regurgitation, education becomes a collaborative venture between students and teachers such that information is constructed rather than simply found (Giroux 1983). Moreover, by leveraging the cultural, social, and human capital of higher education, service-learning practitioners are able to make a visible difference in the communities they are a part of (O'Grady 2000; Cuban and Anderson 2007; Sleeter 2001). This social action creates an opportunity for "border crossing" (Hayes and Cuban 1997) that encourages students, teachers, and community members to question the predominant and hegemonic norms of who controls, defines, and limits access to knowledge and power.

At the same time, a political perspective encourages a reflexive and critical stance toward the foundations and implications of service-learning. It is here that service-learning is found to be a potentially repressive activity, rather than the ameliorative one described from the previous two perspectives. For example, there is little empirical evidence that service-learning provides substantive, meaningful, and long-term solutions for the communities it is supposedly helping. In fact, it may do just the opposite to the extent that it perpetuates and reinforces dominant deficit perspectives of "others" and substantiates the unquestioned norms of whiteness for students engaged in service-learning (Rosenberg 1997; Boyle-Baise 1999; Varlotta 1997a; Sleeter 2001). It is from such a perspective, in fact, that many practitioners and scholars involved in community-based models of teaching and learning balk at the seemingly patronizing label of service-learning as being just that, service (e.g., Schultz 2007).

From such a perspective, service-learning becomes yet another means for those in the "culture of power" to maintain inequitable power relations under the guise of benevolent volunteerism. It reinforces conservative assumptions that relatively isolated actions of caring individuals can overcome societal problems, that it is the servers who bring the solutions, and that such solutions are assimilationist by nature. Tutoring students or working in a soup kitchen maintains the position of privilege for those doing the serving, and presumes that the enactment of such service in and of itself substantiates the worthiness and legitimacy of the servers' perspective.

It is thus inaccurate to portray, as is often done, a tight linkage between a cultural and a political perspective. Service-learning that enhances students' civic responsibility, for example, does not necessarily also develop a stronger democracy. (There are surely numerous historical examples of totalitarian regimes that prided themselves on citizens' sense of civic responsibility.) Civic engagement that leads to political engagement is one option if the engagement mobilizes silenced communities, fosters neighborhood self-reliance, or dramatically increases individuals' aptitudes to understand and participate in our legal, social, and cultural institutions. Yet another option is that service-learning may simply reinforce students' deficit notions that blames the individual or the so-called culture of poverty for the ills that allowed those students to engage in such service in the first place. This is "drive-by volunteerism" at its worst (Cross 2006). A political perspective thus rejects service-learning as an instrumental and amelioristic methodology to focus instead on how service-learning affects power relations among and across diverse individuals, groups, and institutions.

An Antifoundational Perspective on Service-Learning

An antifoundational perspective on service-learning embraces what Dewey (1910) termed a "forked-road" situation of thoughtfulness to foster doubt concerning the normalcy and neutrality of our seemingly commonsensical view of the world. An antifoundational perspective references the philosophical movement of pragmatist antifoundationalism articulated by, among others, Richard Rorty (1989) and Stanley Fish (1985, 1999). This position argues that there is no neutral, objective, or contentless "foundation" by which we can ever know the "truth" unmediated by our particular condition. Fish (1985) argues:

> [Antifoundationalism] is always historicist; that is, its strategy is always . . . to demonstrate that the norms and standards and rules that foundationalist

> theory would oppose to history, convention, and local practice are in every instance a function or extension of history, convention, and local practice. (112)

Antifoundationalism makes us aware of the always contingent character of our presumptions and truths; there is, in Rorty's terminology, no "god's eye view" by which to adjudicate "the truth." Rather, truths are local, contingent, and intersubjective. An antifoundational perspective of service-learning is thus not directed toward some specific and predetermined end goal (such as better comprehension of microeconomics or openness to diversity). It is instead committed to denying us the (seeming) firmness of our commonsensical assumptions. It is, in Dewey's (1910) evocative phrasing, about the need for individuals to "endure suspense and to undergo the trouble of searching . . . to sustain and protract [a] state of doubt" (14, 16) to become thoughtful and educated citizens.

Put otherwise, the end goal of service-learning from an antifoundational perspective is to avoid an all too easily achieved end goal, such as the closing off of an idea or discussion. Service-learning in this vein is about disrupting the unacknowledged binaries that guide much of our day-to-day thinking and acting to open up the possibility that how we originally viewed the world and ourselves may be too simplistic and stereotypical. This condition of possibility for rethinking our taken-for-granted world is what the educational philosopher Gert Biesta (1998) argues is a "radical undecidability" that cannot simply default into an either/or binary. Framed in this light, service-learning allows us to focus as much on the process of undercutting dualistic ways of thinking as on the product of deliberative and sustainable transformational change.

This may be seen as analogous to postmodernism's "incredulity of metanarratives" (Lyotard 1984): there is no single and objective truth to be found, for all perspectives are beholden to particular presuppositions, contexts, and modes of thought. As such, knowledge and meaning become fragmented and partial. While only a few scholars have employed such an antifoundational or postmodern perspective to analyze service-learning (see Boyle-Baise 1999; Harvey 2000; Himley 2004; Varlotta 1997a, 1997b), antifoundational service-learning appears to align itself well with the multifaceted approach supported by a scholarship of engagement committed to interdisciplinarity and the perspectives of multiple stakeholders (Van den Ven 2007). As I detail in Chapter 3, antifoundational service-learning serves as an entrance into what I term "justice learning" from a position of doubt rather than certainty. For example, teacher education students may begin to see that a youth labeled "at-risk" in school may be very different in a tutoring environment, at home, or on an outing; a new perspective on multicultural education may be gained by viewing youths' racial and ethnic groupings, subgroupings, and cross-groupings, as well as by hearing youths' own voices and perspectives; the definitional certainty of what constitutes success and failure may be disrupted in the face of the community strength and vibrancy of marginalized groups and the community's own perspective of what constitutes success.

An antifoundational perspective therefore does not presume that service-learning is a fundamentally positive or negative activity. This is not to say that it reflects a radical relativism (Butin 2001). As Foucault (1997) suggested, the "point is not that everything is bad, but that everything is dangerous . . . so my position leads not to apathy but to a hyper- and pessimistic activism" (256). This perspective informs, as I lay out in Chapter 7, a grounding for supporting faculty's differential appropriation and "buying into" service-learning and community engagement.

MULTIPLE PERSPECTIVES AND VEXING PROBLEMS IN SERVICE-LEARNING

I have previously detailed (Butin 2005a) how such multiple perspectives may inform a rethinking of specific and deep conundrums in the service-learning field (e.g., the limited community impact, the difficulty of enunciating "best practices" that lead to meaningful outcomes, and the difficulty of the rigorous and authentic assessment of such outcomes). Let me as such simply highlight two of these examples to suggest that many issues, above and beyond the ones outlined here, continue to plague service-learning practice and scholarship precisely because of a lack of multivocality in the definitions, criteria, and conceptualizations of what service-learning is and could be.

The first example is of the limited community impact of service-learning. Cruz and Giles (2000) have noted that the "service-learning research literature to date is almost devoid of research that looks at the community either as a dependent or independent variable" (28). They suggest that this lack is due to the theoretical, methodological, and pragmatic difficulties of rigorously defining and analyzing such constructs as "community" and "community impact." Although certain research (e.g., Gelmon et al. 1998) does suggest that positive community impact accrues from service-learning, the lack of empirical research seems to serve as a proxy for the traditional lack of institutional emphasis, practitioner awareness, and community organizations' lack of voice with regard to the role of the community (see, though, the work of Randy Stoeker [e.g., 2002, 2003] as an important development to overcome these exact problems).

Put otherwise, a technical perspective of service-learning simply presumes that a lack of community impact need solely be addressed by enhancing additional technical aspects of the service-learning protocol—for example, by increasing time requirements in the placement site, implementing authentic assessment of community impact, and creating stronger partnerships between educational institutions and the community. Yet it is difficult to address the lack of community impact from a unidimensional perspective. Rather, a technical perspective must be linked to a political perspective to ask questions such as: Who benefits from service-learning practice as traditionally enacted? What are the structures that maintain and allow such benefits to accrue?

In doing so, service-learning becomes revealed to be almost universally located within the context of a specific academic course beholden to specific structural constraints: there is a short-term, one-semester time frame to complete activities; there are a limited number of engaged students; there is a complete turnover of the "service" population; the goals of the course are student-centered to the extent that academic learning is a key requirement within the course; service-learning is positioned as an add-on that can easily be put in or taken out of a course; there is a limit to the time that students and teachers can be involved; and the service-learning on the academic side is ultimately associated with a particular individual (namely the professor supporting the service-learning). These structural conditions make clear that service-learning, as a classroom-based practice, privileges the students (and teachers) in a particular course: they gain knowledge and insight; they participate in a quasi-experimental study on the so-called community for academic gain; they feel good about themselves; they gain peer and institutional approval and recognition; and they gain "real world" experience that can be easily put into a résumé.

Once the privileging of a course-based structure is made clear, it becomes possible to rethink the efficacy of different spaces for service-learning, specifically, at the departmental and institutional level. For if the primary goal for service-learning is truly community impact, then the classroom level, with the structural constraints enumerated above, appears to be the least likely to create an impact. At the departmental level, for example, one could construct linkage and consistency across courses, departmental resources (e.g., administrative support, academic legitimacy) become available, and long-term projects become possible (Berle 2006). At the institutional level, "institutional ownership"—with its access to human, social, and fiscal capital, and its ability to impose and create cultural norms—makes possible an overarching institutional "culture of service." This is the philosophy underpinning the Carnegie Foundation's "community engagement" voluntary classification (Carnegie 2006).

A second issue with a unidimensional perspective is the limited empirical evidence for defining and articulating "best practices" that foster meaningful and substantive student outcomes. Research consistently shows a small but significant increase in academic, social, and personal outcomes due to service-learning. Nevertheless, while researchers have begun to articulate *what* positive outcomes may accrue from service-learning, there is almost no solid research on *how* such outcomes occur.

Reflection, for example, is seen as a key component in service-learning; yet any definition of its duration, scope, placement, mode, and structure remain frustratingly absent. Every teacher of a service-learning course must either implicitly or explicitly decide, among other things, what students should reflect on; how long and how often they should reflect; whether reflection should be in class, out of class, or some combination thereof; what mode of reflection is valid (e.g., monologue, dialogue, performance, written); the level of descriptive, analytic, and reflective detail; and the means by which such reflection will be assessed (e.g., self-, criterion-, or norm-referenced). There is simply no rigorous research of service-learning practice that begins to address this level of detail (I exclude here the proliferation of anecdotal and retrospective self-reporting data).

This issue is, from a technical/cultural perspective, troubling. Without adequate definitions, practitioners cannot develop optimal learning environments for enacting service-learning, researchers cannot rigorously ascertain the "value-added" component of service-learning, and policymakers cannot focus legislative support on "best practices" supported by "scientifically based research." As Furco and Billig (2002) argue, more substantive research is critical to "bring us one step closer to understanding the essence of service-learning" (viii).

Yet an antifoundational perspective argues that there is no such thing as an "essence" of service-learning. There are, instead, truth claims about service-learning that struggle for normative sovereignty and as such produce our identities as service-learning providers or recipients. The pursuit of "best practices," from an antifoundational perspective, is more clearly understood as a contested construction of social and cultural categories by which we define who we are and what we do.

The point for service-learning is that the quest for definitional certainty has the potential to constrain rather than foster emergent practices. An antifoundational perspective suggests that researchers' attempts to pinpoint the *how* of service-learning privilege quantification and thus normalization. For example, to construct a "best practice" for reflection, no matter the "good intentions," is to move the discussion away from the *usefulness* of reflection in multiple modes and arenas to the *legitimacy* of diverse methodologies by which reflection is enacted.

The clearest example of this mentality is the type of criteria developed by institutions to gauge what constitutes a service-learning course. A course typically "becomes" a service-learning course if it requires students to fulfill a certain number of contact hours in the community and offers an opportunity for the student to reflect on the "value" of the experience through some form of "reflection," either a discussion, journal writing, or essay. The singularly normative implications are to develop specific technical standards—e.g., number of contact hours, type of reflection activity—to which practitioners are then held accountable. But in doing so, such practices constrain and disallow alternative and potentially more fruitful means by which to gauge the usefulness of reflection. Research on the role of reflection in service-learning should be less concerned with issues of how long reflection should be done and more with issues of how reflection may better support self-awareness and self-reflective practice.

IMPLICATIONS OF MULTIPLE PERSPECTIVES: SERVICE-LEARNING AS STRATEGY

So if service-learning is not just "a course-based, credit-bearing, educational experience," if it can be viewed and enacted from multiple perspectives, what exactly is it? Let me suggest that it is a strategy. By this I mean that since service-learning can be enacted in multiple ways—technical, cultural, political, or antifoundational—it is a strategic pedagogical and/or philosophical decision. Irrespective of whether such a strategic move is consciously made or not, the means of "doing" service-learning becomes the framework within which to understand the linkage across teaching, learning, and research in the higher education classroom and local community. This is because service-learning is an experiential activity that is always already a culturally saturated, socially consequential, politically contested, and existentially defining experience.

Service-learning is culturally saturated to the same extent that any other complexly enacted process is in our society. It bears the assumptions and implications of our cultural models, for example, those of growth, progress, individualism, and agency. When I interact with colleagues, give out homework assignments, or watch a playground basketball game, I carry with me all of the (often contradictory) cultural assumptions of my particular local, regional, and national affiliations that impacts what I see, how I see it, and how I interpret it. More forcefully, one cannot separate what we think from who we are. The implications are that service-learning must be "read" as any other cultural practice, for there is no transparent, neutral, and objective position by which anyone and everyone understands what we mean when we say that we "do" service-learning.

Service-learning is socially consequential in the sense that all outcomes of the service-learning process, no matter how great or small, have impact in and on the world. For irrespective of the actual scope, duration, or outcome, all service-learning is done *by* individuals *with* other individuals. Whether this is explored through a social justice perspective (Freire 1994) or through a developmental identity framework (Baxter Magolda 1999; Tatum 1992), one must attend to the consequences of service-learning, from the minutiae of individuals' interpretations to the vastness of an entire community's change.

Service-learning is politically contested in that it is fundamentally an attempt to reframe relations of power. This may be thought of in traditional models of empowering bottom-up change or in more postmodern notions of destabilizing foundational assumptions of knowledge, power, and identity. In either case, service-learning promotes a deviation from the status quo, if only because (at minimum) it just does things differently. On a deeper level, service-learning disturbs our society's penchant for security, order, and control, all of which are presumed to be synonymous with "safety." But as articulated above, service-learning is not safe. It is anything but safe. As such, all interventions that promote such disturbances—to the individual, the institution, or the community—are deemed political and thus contested.

Finally, service-learning is existentially defining because it forces individuals (students, faculty, and community partners) to take a stance. In doing so, individuals must (consciously or not) define themselves by the decisions they make or refuse to make. One cannot remain neutral when engaging in service-learning. Even the attempt to remain so positions oneself in a particular resistant identity, for it becomes clear that service-learning pulls strongly at the strings that bind us and support us. And in that pulling and pushing, we must as existential beings decide in which direction we will be moved. Do I intervene in an argument between two youth? Do I stand up and speak at the local neighborhood meeting? Do I reveal to my community partner that I may not be here next semester? Every action—even when the decision not to act is taken—reveals and defines (to a lesser or greater extent) who we are and what we believe.

This is why service-learning is a strategy, specifically, a strategy of disturbance. By this I mean that service-learning challenges and decenters our static and singular notions of teaching, learning, and research by moving against the grain of traditional practice in higher education: that is, it is a deeply engaging, local, and impactful practice. This is the "disturbing" part: service-learning is a pedagogical practice and theoretical orientation that provokes us to more carefully examine, rethink, and reenact the visions, policies, and practices of our classrooms and educational institutions; it forces us—as faculty, administrators, and policymakers—to confront the assumptions and implications of the ways in which we teach and learn and conduct scholarship.

And it is a strategy, a pedagogical strategy, because it is a conscious intervention into local and highly complex contexts. Foucault (1997) suggested that one can never escape relations of power and "regimes of truth"; rather, "one escaped from a domination of truth not by playing a game that was totally different from the game of truth but by playing the same game differently, or playing another game, another hand, with other trump cards . . . by showing its consequences, by pointing out that there are other reasonable options" (295–96). A situation that we once took for granted is revealed as fundamentally different once a new goal becomes visible or a hidden presumption is revealed. So if service-learning can reveal the limits of traditional models of teaching, learning, and research, it serves its purpose as a set of "trump cards" to be played.

Put otherwise, service-learning makes us take a stand by acting up and acting out, irrespective of the lens within which we operate. It is much easier to teach within the boundaries of the normal. While I do not suggest that lecture-hall or seminar-style teaching is "easy," I do suggest that such teaching sidesteps and thus suppresses fundamental questions of higher education pedagogy: How is knowledge created and by whom? What is the "usefulness," if any, of disciplinary knowledge? What is the role of higher education in a liberal democracy? What is the role, moreover, of students, faculty, and institutions in their local and global communities? While the answers will obviously differ across institutions, the questions do not. Put otherwise, the normative silence on pedagogical practice by individual faculty and higher education institutions promotes and perpetuates traditional models of teaching and learning. This privileges top-down presumptions of knowledge transfer from faculty to students and power relations between institutions and community and institutions and faculty. By implementing powerful service-learning programs, individuals must act up the institutional hierarchy.

Likewise, acting out—outside of traditional departments, outside of physical classroom walls, outside of the proximity and "safety" of the academic campus—is a disturbing endeavor on pragmatic, political, and existential grounds. It is extremely difficult to pragmatically implement a powerful service-learning program. It takes foresight, time, organizational capabilities, creativity, networking skills, tolerance for ambiguity, willingness to cede sole control of classroom learning, and an acceptance of long-term rather than immediate increments of progress. It takes convincing—oneself and others—that the boundaries of academic disciplines, classroom walls, and institutional boundaries are socially constructed and thus changeable. Yet, pragmatically speaking, such social constructions create normative pressures, solid walls, and clear institutional structures, all of which must be circumvented or worked through in order to implement service-learning programs.

Service-learning is also politically disturbing for individual faculty. It is a practice that might not be rewarded by traditional tenure and promotion guidelines, that questions (either implicitly or explicitly) colleagues' pedagogical practices, and that has the potential to turn out badly in a very public and glaring way. Given the high-stakes nature of the tenure review process, engaging in a nontraditional methodology is a disheartening proposition for new and junior faculty.

Finally, service-learning is an existentially disturbing endeavor. By this I mean more than an individual's necessary fortitude and courage to confront and overcome the pragmatic and political obstacles to implementing service-learning. I mean that service-learning, when deeply done, subverts some of our dearest foundational assumptions of our sense of identity as higher education faculty. We must rethink the belief that academic knowledge comes directly from us, in a classroom, based on a written text, and assessed objectively. We must acknowledge our students as active, reflective, and resistant agents in their own educational processes. We must come to terms with the reality that our particular expertise may have very little currency (or even relevance) in the messy and complex world outside our classroom walls.

It may be fruitful to view an analogous field to understand the value of multiple and conflicting perspectives on service-learning and its destabilizing potential. Multicultural education has spent the last thirty years grappling with and developing the distinctions it is premised on (Sleeter and Grant 2003; Banks 1996). The civil rights movements of the 1960s and '70s shattered the ameliora-tory presumptions of a paternalistic and assimilationist "melting pot" view of educational practices. Instead, multicultural educators developed a host of critical theories premised on fundamentally different assumptions of what counts as multicultural education: ethnic studies perspectives focused on educational practices and norms aligned to the cultural perspectives of those being taught (e.g., queer studies, women's studies, Afrocentric schooling [Asante 1998; Rich 1979]); difference multiculturalists emphasized the diversity of means by which we could think of learning, intelligence, and success (e.g., multiple intelligence [Gardner 1983]); and critical multiculturalists demanded that issues of equity and social justice be at the heart of educational practices (e.g., problem-posing education [Freire 1994]).

Multicultural educators have thus put forward multiple alternative articulations in attempts to rethink and reframe what multicultural education should be premised on and moving toward. Every theoretical strand can be seen as a specific response to specific pedagogical or political problems (e.g., continued lack of equitable outcomes; lack of cultural congruence between students and textbooks).

What is glaring in the service-learning field vis-à-vis multicultural education is the lack of analogous articulations of multiple and distinctive foundational assumptions. Such overreliance on a singular vision, I suggest, discourages a rigorous analysis and critique of the foundational terminology of service-learning for fear of a loss of meaning. Yet without multiple (and competing) foundational premises, the field is beholden to embracing potentially pragmatically limited and theoretically problematic articulations of service-learning.

If service-learning is to avoid becoming overly normalized, we must continuously question and disturb our assumptions, our terms, and our practices. I do this in the next two chapters as a means of rethinking and thus rearticulating service-learning as a form of "justice learning." But one cannot do this solely by claiming that service-learning is, in and of itself, a type of justice-oriented education. One must clear away the underbrush of existing assumptions to redevelop a new vision.

The next two chapters, as such, disturb the normalizations in place within the service-learning field. Not to do away with them so much as, again following Foucault here, to demonstrate their consequences and how other options may play out. By working through other modes of service-learning in other ways, I hope to open up the field for further questioning and experimentation.

Chapter 2

The Limits of Service-Learning

In this chapter I want to explore in more depth the unacknowledged and unanticipated problematics of service-learning in higher education, particularly as service-learning practice and theory has turned in the last few years to ensuring its institutional longevity. This chapter thus takes a critical look at this attempted institutionalization of service-learning in higher education. It asks whether service-learning can become deeply embedded within the academy, and, if so, what exactly is it that becomes embedded. Specifically, I suggest that there are substantial pedagogical, political, and institutional limits to service-learning across the academy. These limits, moreover, are inherent to the service-learning movement as contemporarily theorized and enacted. As such, there may be a fundamental and unbridgeable gap between the rhetoric and reality of the aspirations of the present-day service-learning movement.

I first situate the service-learning movement through an analysis of its drive toward institutionalization. Such an analysis reveals some of the fundamental and underlying assumptions of the service-learning field. I then proceed to show how these assumptions harbor significant pedagogical, political, and institutional impediments against the authentic institutionalization of service-learning. I conclude this chapter by suggesting how a reframing of such assumptions may allow service-learning to be repositioned as a disciplinary field more suited to becoming deeply embedded within higher education, which I then explore in the second part of this book.

Institutionalizing Service-Learning

After a heady decade of growth, the service-learning movement appears ideally situated within higher education. It is used by a substantial number of faculty across an increasingly diverse range of academic courses; administrative offices and centers are devoted to promoting its use; it is prominently cited in presidents' speeches, on institutional homepages, and in marketing brochures. Yet as the latest Wingspread statement (Brukardt et al. 2004) put it: "The honeymoon period for engagement is over; the difficult task of creating a lasting commitment has begun" (4). For underneath the surface, the service-learning movement has found its institutionalization within higher education far from secure. As a recent report (Saltmarch, Hartley and Clayton 2009) has noted, the civic engagement movement has stalled, due, in part, to its being inadequately conceptualized and highly fragmented; it verges on "stand[ing] for anything and therefore nothing" (4).

I have already outlined many of the pragmatic issues: for example, the lack of institutionalized budget line items, lack of interest from key and tenured faculty, and consistent reconstruction of service-learning practices and policies by even so-called stable and well-functioning programs. More troubling still is that the academy's "buying into" service-learning may be much easier said than done, with little political or institutional costs for failing to achieve substantial goals. Rhetoric may be winning over reality. It is thus that the Wingspread participants (2004, ii) "call the question": "Is higher education ready to commit to engagement?" This can be framed in poker parlance as "calling the bluff." Does higher education have the desire, the long-term fortitude, and the resources to remake itself? Is higher education able, for the sake of itself, its students, and American society more generally, to embrace a more engaged, democratic, and transformative vision of what it should be, should have been, and was before (Benson et al. 2005; Harkavy 2006)? If so, then it better ante up.

There is thus a burgeoning literature on the institutionalization of service-learning. I want to focus on Andy Furco's work (2001, 2002, 2003; Furco and Billig 2002) and on the Wingspread statement because each takes a diametrically opposed stance on the means of institutionalizing service-learning; both, though, carry exactly the same presumption of what the outcomes of such institutionalization should be—an overarching "meta" framework by which to recreate higher education as an engaged and civically minded institution. While the literature is ever growing and far from singular in perspective

(Bell et al. 2000; Benson et al. 2005; Bringle and Hatcher 2000; Gray et al. 2000; Hartley et al. 2005; Holland 2001; Kramer 2000), Furco's work and the Wingspread statement are emblematic of this dominant vision and the goals of service-learning institutionalization and the two primary and divergent paths to achieving such goals. Specifically, Furco's work offers a systematic rubric for gauging the *incremental* progress of service-learning institutionalization; the Wingspread statement, on the other hand, promotes a *transformational* vision for service-learning in higher education.

The educational historian Larry Cuban (1990, 1998) has cogently referred to this distinction as first- versus second-order change and has explored the historical contexts and conditions that support one form of educational reform over another. Of interest here is that irrespective of the divergent means propounded, both perspectives have a vision of service-learning as a metatext for the policies, practices, and philosophies of higher education. Thus irrespective of how it is to be institutionalized, service-learning appears as the skeleton key for unlocking the power and potential of postsecondary education as a force for democracy and social justice. This is what I talk about later in the book when I focus on the notion of service-learning as a social movement. By further explicating the divergent means propounded by incrementalist and transformationalist perspectives, it becomes possible to grasp the overarching assumptions and implications of the service-learning movement.

Furco (2002) has developed a rubric for viewing the institutionalization of service-learning. The rubric works as a road map that may be followed by individuals and institutions committed to embedding service-learning throughout their campuses, and it works as a formal or informal assessment mechanism to gauge the progress along the institutionalization path. Furco operationalizes institutionalization across five distinct dimensions "which are considered by most service-learning experts to be key factors for higher education service-learning institutionalization" (1): (1) philosophy and mission; (2) faculty support and involvement; (3) student support and involvement; (4) community participation and partnerships; and (5) institutional support. While Furco argues elsewhere (2001, 2003) that research shows faculty and institutional support to be the key institutional factors, the rubric makes clear that "what is most important is the overall status of the campus' institutionalization progress rather than the progress of individual components" (3).

The real value and usefulness of the rubric is that it clearly and succinctly lays out the step-by-step increments by which a campus can institutionalize service-learning. Faculty support for and involvement in service-learning, for example, moves from "very few" to "an adequate number" to "a substantial number" of faculty who are knowledgeable about, involved in, and leaders of service-learning on a campus. Staffing moves from "no staff" to "an appropriate number . . . paid from soft money or external grant funds" to "an appropriate number of permanent staff members" (13). The rubric does not suggest how such incremental progress is to be achieved; each campus culture and context is different. Instead, it lays out an explicit framework for (in Cuban's terminology) "tinkering" toward institutionalization.

The Wingspread statement (Brukardt et al. 2004) has a fundamentally different agenda: "Our goal in calling the question is nothing less than the transformation of our nation's colleges and universities" (ii). To accomplish this goal the statement articulates six specific practices to institutionalize engagement: (1) integrate engagement into mission; (2) forge partnerships as the overarching framework; (3) renew and redefine discovery and scholarship; (4) integrate engagement into teaching and learning; (5) recruit and support new champions; and (6) create radical institutional change. Many of these practices mirror Furco's rubric and can possibly be implemented without radical transformation: integrating engagement into a mission statement, forging stronger partnerships, fostering more engaged pedagogy, and recruiting new voices to speak for engagement are all doable without fundamentally altering the structure and practices of higher education.

Of course such changes, if truly and deeply implemented, *should not* be doable without fundamentally altering the structures and practices of higher education. The Wingspread statement is premised exactly on the notion that these practices would be taken up in "thick" ways that would force the restructuring of much of how we "do" higher education. Unfortunately, these practices, as articulated, are all too easily misappropriated within the world of higher education. This is not to suggest that these practices are not important. In fact, they may actually be the most sustainable aspects of service-learning as presently conceived. The point here is simply that they are not at the heart of what the Wingspread statement really *means* when it talks about institutionalizing service-learning.

What it really means, and what the third and sixth practices make vivid as truly radical suggestions, is the desire to transform higher education through service-learning. Redefining scholarship and creating radical institutional change by, for example, overturning higher education's "hierarchical, elitist and competitive environment" (15) is a revolutionary call to arms. And the Wingspread statement is well aware of this: each specific practice has a "What is needed" section that offers concrete action steps (e.g., "expanded assessment and portfolio review options for faculty" [14] to integrate engagement into teaching and learning); under the "Create radical institutional change" section, what is needed is "courage!" "new models," "serious . . . funding," "new links between academic work and critical public issues," and "institutional flexibility and willingness to experiment—and to fail." These are not action steps. This is a battle cry.

Thus where Furco's rubric offers a deliberate and deliberative procession of rational increments, the Wingspread statement and its later rearticulations (see, e.g., Saltmarsh et al. 2009) provide a fiery manifesto for reinvention. Irrespective of which model is better (or if perhaps both are necessary), what is of interest is that both presume that, by whichever means necessary, service-learning should become an overarching framework for higher education. This framework, moreover, should be embedded both horizontally across departments and vertically throughout all levels of an institution's pronouncements, policies, and practices. Both presume that service-learning can and should be done in departments from Accounting to Women's Studies; that all students, faculty, administrators, and community partners can be involved; and that everything from line-item budgets to institutional Web pages have the imprint of service-learning.

This is nothing less than a grand narrative for higher-education-as-service-learning, for it thinks about service-learning as a politics to transform higher education and society. The implications are both prominent and problematic. Such a perspective presumes that service-learning is a universal, coherent, cohesive, amelioratory, and liberatory practice. It presumes that service-learning is not somehow always already a part of the institutional practices and norms it is attempting to modify and overcome. Yet, as I argue below, such presumptions are unfounded. In fact, thinking about service-learning as a politics of transformation leads to a theoretical and political cul-de-sac.

THE LIMITS OF INSTITUTIONALIZATION

This section questions the notion of service-learning as an overarching and transformative agent of social change and social justice within higher education and society more generally by focusing on three specific claims made by the service-learning movement: service-learning as a means (1) to transform pedagogy; (2) to usher in a more democratic and socially just politics within higher education; and

(3) to redirect postsecondary institutions outward toward public work rather than inward toward academic elitism.

These claims, it should be noted, are premised on an inherent compatibility between service-learning and the academy. This seeming compatibility indexes presumptions that civic engagement and "real world" learning are hallmarks of the future of higher education. Yet such presumptions are of course open to contestation and critique, perhaps the most biting of which has come from Stanley Fish.

Fish (2004a, 2008) has opined that we should stick to questions about the truth and not bother with issues of morality, democracy, or social justice: "We should look to the practices in our own shop, narrowly conceived, before we set out to alter the entire world by forming moral character, or fashioning democratic citizens, or combating globalization, or embracing globalization, or anything else" (Fish 2004a, A23). Fish was responding directly to a publication from a group of scholars at the Carnegie Foundation's Higher Education and Development of Moral and Civic Responsibility Project (Colby et al. 2003), but his critique has general resonance for those who see the academy as primarily a site of knowledge production and dissemination rather than of something as nondefinable and potentially partisan as moral and civic betterment.

I am sympathetic to Fish's arguments and address them in the next chapter (see also Butin 2008). The real issue, though, and as the second part of this book details, the question today is no longer *if* service-learning is to become a part of the academy so much as *how* it is already becoming a part of it and the implications thereof. It is thus necessary to work through the pedagogical, political, and institutional limits that are too often unacknowledged and unanticipated in traditional service-learning theories and practices.

Pedagogical Limits to Service-Learning

Service-learning, as I have argued, is seen as a transformative pedagogy that links classrooms with the real world, the cognitive with the affective, and theory with practice to disrupt the banking model of education that presumes passive students ingest neutral knowledge from expert faculty, and then regurgitate such knowledge in discrete and quantifiable measures (Freire 1994; hooks 1994).

But is this transformation possible? Campus Compact's (2004) annual membership survey shows the following departments with the highest offering of service-learning courses: Education (69 percent), Sociology (56 percent), English (55 percent), Psychology (55 percent),

Business/Accounting (46 percent), Communications (46 percent), and Health / Health Related (45 percent). In a now-classic formulation, Tony Becher (Becher and Trowler 2001) argued that academic disciplines can be differentiated along two spectra: "hard/soft" and "pure/applied." "Hard-pure" fields (e.g., chemistry and physics) view knowledge as cumulative and are concerned with universals, simplification, and quantification. "Hard-applied" fields (e.g., engineering) make use of "hard-pure" knowledge to develop products and techniques. "Soft-pure" fields (e.g., English) view knowledge as iterative and are concerned with particularity and qualitative inquiry. "Soft-applied" fields (e.g., education, management) make use of "soft-pure" knowledge to develop protocols and heuristics. What becomes immediately clear is that service-learning is overwhelmingly used in the "soft" disciplines. Biology is the highest "hard" field (at number 10 with 37 percent), with the natural sciences next (at number 18 with 25 percent).

It should of course be acknowledged that the "hard/soft" and "pure/applied" distinctions are socially constructed monikers that carry long-standing ideological baggage and serve as proxies for contestations surrounding the power, legitimacy, and prestige of any particular discipline. Scholars in the sociology of knowledge and history of science have shown not simply the ambiguity and permeability of the boundaries between so-called soft and hard disciplines, but have fundamentally questioned the (to use Foucault's terminology) "scientificity" of claims to the objective and neutral practice of mapping reality (Hacking 1999; Lather 2005; Latour 1979). Yet what is at issue here is not whether there is "really" a distinction between the "hard" and "soft" sciences, but how such a socially constructed distinction is ultimately determined and practiced in our day-to-day life. As Cornel West (1994) once wryly noted, taxicabs in Harlem still did not stop for him even if race was a social construction. Likewise, there is a plethora of empirical evidence (Biglan 1973; Lueddeke 2003; NCES 2002) that teaching practices differ significantly across disciplines; as such, these disciplinary distinctions serve as useful heuristics for understanding how service-learning may or may not be taken up across the academy.

The service-learning field acknowledges that "soft" disciplines are much more apt for making use of service-learning. Yet the presumption is that this fact is simply a consequence of *either* the poor marketing of what service-learning can offer to the "hard" sciences (from an incrementalist perspective) or the inability of the "hard" sciences to transform themselves into useful public disciplines (from a transformational perspective). What both of these perspectives miss is that Becher's typology demonstrates that each grouping of disciplines manifests "its own epistemological characteristics . . . [of] curriculum, assessment and main cognitive purpose . . . [and] the group characteristics of teachers, the types of teaching methods involved and the learning requirements of students" (Neumann and Becher 2002, 406).

Of most salience here are the divergent notions of teaching styles and assessment procedures between "hard" and "soft" disciplines. I will focus only on the "hard" disciplines here to make vivid its antipathy to service-learning presumptions. Given the sequential and factual nature of the "hard" disciplines, lecturing is predominant. Moreover, the cumulative nature of knowledge makes moot any notion of student perspectives or "voice" in the field. It is simply not relevant how students "feel" about subatomic particles. As such, "in keeping with their atomistic structure [hard pure knowledge fields] prefer specific and closely focused examination questions to broader, essay-type assignments" (Neumann and Becher 2002, 408). "Objective" tests, norm-referenced grading, and the absence of rubrics (given the right/wrong nature of what constitutes knowledge) are typical.

U.S. Department of Education statistics support these theoretical insights. The most recent available data (NCES 2002, table 16) show that apprenticeships and fieldwork are used much more often by the social sciences and humanities (10–15 percent depending on the discipline) than by the natural sciences (2–3 percent). Humanities and social sciences faculty are almost twice as likely to use research papers than natural science faculty (70–85 percent versus 40–50 percent, respectively), and half as likely to grade on a curve (20–30 percent versus 40–50 percent) (table 18, table 22). While some of this data is confounded by the type of institution (e.g., doctoral versus nondoctoral institutions), fairly distinct patterns and differences between disciplines are visible.

Above and beyond these disciplinary differences, though, emerges a more troubling realization. Fully 83 percent of all faculty use lecturing as the primary instructional method in college classrooms, and this percentage does not drastically change across the type of institution, the rank or tenure status of the faculty, or across disciplines (NCES 2002, tables 15 and 16). (It should be noted that there are numerous methodological ambiguities with the NCES data: the lack of a distinctive service-learning category may obscure its use, the lack of Likert scales may distort the actual use of instructional methods, the conflation of lecturing with discussion, etc.; yet the primary point—the unambiguous marginality of nonlecturing pedagogical methods across higher education—still fundamentally stands.)

Thus irrespective of disciplinary and epistemological differences, the vast majority of faculty across higher education see themselves as embodying the normative (read: non service-learning-oriented) model of teaching and learning. This is further exacerbated by the reality that nontenure track faculty by now constitute more than half of all teaching faculty in higher education (Snyder et al. 2004; Schuster and Finkelstein 2007). A normative model of teaching is thus reinforced by the marginal and transitory status of faculty. There thus appears to be a very low upper limit to the use of service-learning across numerous disciplines and amongst faculty in higher education.

If faculty instructional practices and demographics do not conform to who should make use of service-learning, then student demographics do not align with the type of students supposedly doing service-learning. As I noted in the previous chapter, the service-learning field presumes an "ideal type" service-learning student—one who volunteers her time, has high cultural capital, and gains from contact with the "other." The service-learning literature is replete with discussions of how students come to better understand themselves, cultural differences, and social justice through service-learning. The overarching presumption is that the students doing the service-learning are white, sheltered, middle-class, single, without children, unindebted, and between the ages of eighteen and twenty-four. But that is not the demographics of higher education today; much less will it be so in twenty years.

NCES (Snyder et al. 2004) data show that the largest growth in postsecondary enrollment will be in for-profit and two-year institutions; already today, fully 39 percent of all postsecondary enrollment is in two-year institutions (table 178). Moreover, 34 percent of undergraduates are over twenty-five years of age; 40 percent of undergraduates are part-time. Even considering just full-time undergraduates, more than 18 percent are over twenty-five years of age (tables 176, 177). Additionally, college completion rates continue to be low: less than half of all college entrants ultimately complete a baccalaureate degree, with graduation percentages dipping much lower for two-year institutions and among part-time, lower-class, and/or nonwhite students. Finally, U.S. Census data (U.S. Census 2008) forecasts that white youth will become a numeric minority in our K-12 schools within a generation; this changing demographic wave is already impacting the makeup of higher education.

These statistics raise three serious pedagogical issues for the service-learning field. First, service-learning is premised on full-time, single, nonindebted, and childless students pursuing a "liberal arts education." Yet a large proportion of the postsecondary population of today, and increasingly of the future, views higher education as a part-time, instrumental, and preprofessional endeavor that must be juggled with children, family time, and earning a living wage. Service-learning may be a luxury that many students cannot afford, be it in terms of time, finances, or job future.

Second, service-learning is premised on fostering "border-crossing" across categories of race, ethnicity, class, (im)migrant status, language, and (dis)ability. Yet what happens when the postsecondary population *is already* of those identity categories? The service-learning field is only now beginning to explore such theoretical and pragmatic dilemmas (e.g., Henry 2005; Swaminathan 2005), and these investigations are already disrupting some of the most basic categories within the service-learning field (e.g., the server/served binary; student/teacher and classroom/community power dynamics; reciprocity).

Third, there is the distinct possibility that service-learning may ultimately come to be viewed as the "whitest of the white" enclave of postsecondary education. Given the changing demographics, and given the rise of the "client-centered" postsecondary institution, service-learning may come to signify a luxury available only to the privileged few. Educational research (e.g., Peske and Haycock 2006) has clearly shown how inequities across K-12 academic tracks (e.g., teacher quality, adequate resources, engaging curricula) correlates to youth's skin color and socioeconomic status. Such hierarchies within service-learning in higher education are not unthinkable.

Arguments can of course be made by both incrementalist and transformationalist perspectives. The former will argue that these issues will simply take more time to work through, while the latter will argue that in transforming higher education such issues will become irrelevant. Perhaps. The goal here is not to be defeatist, presentist, or conservative; it is not to argue that higher education is a static and unchangeable beast. Rather, it is simply to map out the structures and norms that inhibit the institutionalization of a viable and powerful service-learning pedagogy.

Political Limits to Service-Learning

Yet even if service-learning is to, in one way or another, overcome the pedagogical barriers just spoken of, what exactly is it that will become institutionalized? By framing service-learning as a politics, advocates may in fact be undermining their most valued goal. Specifically, by viewing service-learning as a universal transformative practice, advocates may allow it to become misappropriated and drained of its transformative potential.

Service-learning has a progressive and liberal agenda under the guise of a universalistic practice. The *Presidents' Declaration on the Civic Responsibility of Higher Education* (Campus Compact 2000), for example, declares: "Higher education is uniquely positioned to help Americans understand the histories and contours of our present challenges as a diverse democracy. It is also uniquely positioned to help both students and our communities to explore new ways of fulfilling the promise of justice and dignity for all . . . We know that pluralism is a source of strength and vitality that will enrich our students' education and help them learn both to respect difference and to work together for the common good" (1). This is a noble and neutral sounding statement. Who could be against "the common good"? Yet what is clear is that the "diversity" and "dignity" being spoken of is not about political conservatives. It is about the multiple populations within the United States who have suffered historically (and many who still suffer today) due to social, cultural, economic, and educational marginalization, degradation, and destruction.

This has a certain natural-seeming quality within the academy, as higher education is supposed to open one's perspective to think and act differently about becoming a public citizen. Yet while this also has a deep resonance with the service-learning field (and some might say it is at the heart of the service-learning field [see Stanton et al. 1999]), it is certainly not the norm in our highly divided red-state / blue-state America. The most obvious example of this is David Horowitz's "academic bill of rights" (http://www.studentsforacademicfreedom.org/documents/1925/abor.html; accessed on November 16, 2009).

Horowitz, the president of the Center for Study of Popular Culture, has crafted a seemingly neutral policy declaration that demands that colleges and universities not discriminate against political or religious orientations such that "academic freedom and intellectual diversity" can flourish in the academy. "Academic freedom," the document states,

> consists in protecting the intellectual independence of professors, researchers and students in pursuit of knowledge and the expression of ideas from interference by legislatures or authorities within the institution itself. This means that no political, ideological or religious orthodoxy

> will be imposed on professors and researchers through the hiring or tenure or termination process, or through any other administrative means by the academic institution. (1)

The document goes on to enumerate numerous principles and procedures that flow from this statement of principle. These include, among others, that faculty cannot be "hired or fired or denied promotion or tenure on the basis of his or her political or religious beliefs," that "students will be graded solely on the basis of their reasoned answers [. . .]," and that "exposing students to the spectrum of significant scholarly viewpoints on the subjects examined in their courses is a major responsibility of faculty. Faculty will not use their courses for the purpose of political, ideological, religious or antireligious indoctrination" (2). This sounds eminently reasonable until one realizes that Horowitz is deliberately attempting to dismantle what he sees as the liberal orthodoxy permeating higher education.

Horowitz (http://www.studentsforacademicfreedom.org/reports/lackdiversity.html; accessed on August 17, 2008) has shown and social science research confirms (Klein and Stern 2005; Rothman et al. 2005) that higher education faculty are overwhelming registered as Democrats, with (according to his data) an overall ratio of 10:1 across departments and upper-level administrations. On some campuses (e.g., Williams, Oberlin, Haverford), Horowitz could not find a single registered Republican faculty member. This, Horowitz (2003) argues, is not diversity: "What is knowledge if it is thoroughly one-sided, or intellectual freedom if it is only freedom to conform? And what is a 'liberal education,' if one point of view is for all intents and purposes excluded from the classroom? How can students get a good education, if they are only being told one side of the story? The answer is they can't" (1). The attack on the liberal bias in higher education is not new. What is new, though, is the strategies propounded by Horowitz (2003):

> I have undertaken the task of organizing conservative students myself and urging them to protest a situation that has become intolerable. *I encourage them to use the language that the left has deployed so effectively in behalf of its own agendas.* Radical professors have created a "hostile learning environment" for conservative students. There is a lack of "intellectual diversity" on college faculties and in academic classrooms. The conservative viewpoint is "under-represented" in the curriculum and on its reading lists. The university should be an "inclusive" and intellectually "diverse" community. I have encouraged students to demand that their schools adopt an "academic bill of rights"

> that stresses intellectual diversity, that demands balance in their reading lists, that recognizes that political partisanship by professors in the classroom is an abuse of students' academic freedom, that the inequity in funding of student organizations and visiting speakers is unacceptable, and that a learning environment hostile to conservatives is unacceptable. (2–3; emphasis added)

Service-learning is not explicitly on the list of Horowitz's grievances. It very well could be. The service-learning literature is replete with students' resistance to the implicit and/or explicit social justice emphasis. Susan Jones (2002; Jones et al. 2005), for example, has carefully shown how student resistance manifests itself in service-learning experiences and how instructors might—through a "critical developmental lens"—begin to overcome such resistance. Yet what is clear is that this is not about liberals resisting a conservative agenda; as one resistant student wrote in Jones's end-of-semester evaluation: "I don't enjoy the preaching of a debatable agenda in the first hour. Perhaps teaching from a more balanced perspective would be better than 'isms are keeping us down.' [. . .] More emphasis on community service. Less on ideologically driven readings and lessons" (Jones et al. 2005, 14).

The point is not that service-learning should stop having an ideological agenda, nor that service-learning should now embrace conservative service-learning to provide "balance." Rather, it is that service-learning embodies a liberal agenda under the guise of a universalistic garb; it is, to put it bluntly, ripe for conservative appropriation.

To date, for example (according to Students for Academic Freedom [www.studentsforacademicfreedom.org], an organization affiliated with Horowitz that tracks such measures), over a dozen states' legislatures (including California, Indiana, Florida, Ohio, Tennessee, North Carolina, New York, and Pennsylvania) have proposed legislation patterned on the "academic bill of rights." Numerous higher education organizations and consortiums of leading higher education institutions have, in response, released their own response of what constitutes academic freedom (e.g., ACE 2005; AAUP 2007). Such beliefs have, moreover, filtered into the consciousness of the general population, as a recent Gallup poll (Wilson 2008) shows that 40 percent of respondents believed that professors oftentimes used the classroom as a platform for their personal political views. An era of legislative and public scrutinizing of higher education's political practices has begun.

Horowitz (or any university president under public pressure) can thus very easily raise the specter of service-learning offices indoctrinating first-year students into biased, unscientific, and undefendable liberal groupthink practices through, for example, daylong conferences about capital punishment or women's rights. The solution? Horowitz would argue that either the entire service-learning office needs to be dismantled to avoid such blatant political abuse of public funds or that the university needs to completely rethink and redo how it helps students to think about such issues; by allowing undergraduates to work, for example, with a pro-life group to send out mail or to picket with a retentionist organization committed to keeping the death penalty.

Service-learning is in a double bind. If it attempts to be a truly radical and transformative (liberal) practice, it faces potential censure and sanction. If it attempts to be politically balanced to avoid such an attack, it risks losing any power to make a difference. At the root of this double bind and the reason it cannot escape from this dilemma is that service-learning has positioned itself as a universalistic and thus neutral practice.

But as Stanley Fish (1999) has pointed out, there is no such thing. "If, for example, I say 'Let's be fair,' you won't know what I mean unless I've specified the background conditions in relation to which fairness has an operational sense" (3). No statements or positions are value-free; they come saturated with particular historical, social, and cultural baggage. Thus not only do genuinely neutral principles not exist, when seemingly neutral principles *are* articulated, it is a blatantly political and strategic move:

> Indeed, it is crucial that neutral principles not exist if they are to perform the function I have described, the function of facilitating the efforts of partisan agents to attach an honorific vocabulary to their agendas. For the effort to succeed, the vocabulary (of "fairness," "merit," "neutrality," "impartiality," mutual respect," and so on) must be empty, have no traction or bite of its own, and thus be an unoccupied vessel waiting to be filled by whoever gets to it first or with the most persuasive force. (7)

Seemingly neutral principles are thus used strategically to promote one's specific ideological agenda, irrespective of political orientation. This is exactly what Horowitz has done with "intellectual diversity" and what the service-learning movement is attempting to do with "civic engagement." But in attempting to hold the (imaginary) center, such strategizing in fact politicizes the term in question through binary extremism. In the former case, "intellectual diversity" becomes a stalking horse for rightwing conservatism; in the latter case, "civic engagement" becomes linked to radical left-wing demands for "social justice."

Service-learning thus finds itself positioned as attempting to deliver a very specific and highly political notion of the truth under the guise of a neutral pedagogy. Its overarching stage theory of moving individuals and institutions from charity-based perspectives to justice-oriented ones in fact maps directly onto our folk theories of what constitutes Republican and Democrat political positions: Republicans believe in individual responsibility and charity, while Democrats focus on institutional structures and social justice (Westheimer and Kahne 2004).

To claim service-learning as a universalistic practice available to all political persuasions is thus to ignore its politically liberal trappings as presently conceptualized and enacted. To cite just one obvious counterexample, is it service-learning if Jerry Fallwell's Liberty University requires a graduation requirement that all undergraduates go out and help blockade abortion clinics to save the lives of the unborn? What if this activity was linked to reflection groups and learning circles and students had to create portfolios showing how such community service was linked to their academic courses?

Few service-learning advocates, I suggest, would quickly or easily accept that this is service-learning, much less a service-learning committed to social justice. But to not accept such a counterexample is to admit that service-learning is not a universalistic practice. It is to admit that service-learning is an ideologically driven practice. And in doing so, service-learning falls exactly into the "intellectual diversity" trap. And once trapped, there is no way out; service-learning, to survive in higher education, will have to become "balanced."

Institutional Limits to Service-Learning

I have suggested so far that service-learning faces major pedagogical and political barriers to becoming institutionalized. Yet if service-learning could overcome these pedagogical and political barriers, would it then be the truly transformative movement envisioned? Sadly again, I doubt it, for higher education works by very specific disciplinary rules about knowledge production, who has the academic legitimacy to produce such knowledge, and how such knowledge is produced (Messer-Davidow et al. 1993). The very institution that service-learning advocates are trying to storm, in other words, may drown them.

The clearest example of this already ongoing process is what I'll term the "quantitative move" in the service-learning field. Put otherwise, service-learning scholarship is becoming adept at using the "statistically significant" nomenclature. The idea is to show that service-learning can, holding all other variables constant, positively impact student outcomes. Thus a wide variety of scholarship has shown service-learning to be statistically significant in impacting, among other things, students' personal and interpersonal development, stereotype reduction, sense of citizenship, and academic learning (see Eyler et al. 2001, for a comprehensive summary). Much of this research has very low *betas* (i.e., the actual impact is not, statistically speaking, profound); nevertheless, service-learning has been "proven" to make a measurable difference in a positive direction vis-à-vis other pedagogical and institutional variables.

The idea behind this quantitative move is obvious: service-learning advocates want to show that service-learning is a legitimate practice with legitimate, consequential, and measurable outcomes in higher education. When in Rome, the thinking goes, do as the Romans do. The problem is that Rome has burned; there are three distinct reasons why the quantitative move ultimately will not help to institutionalize the kind of service-learning hoped for.

The first reason is that quantifying the value-added component of service-learning is methodologically impossible: there are simply too many variables commingling and interacting with each other to allow for valid and reliable conclusions. The number of variables, from types of sites to types of interactions to types of reflections to types of teaching styles, becomes unmanageable for accurate quantification and measurement. In this way service-learning is analogous to teaching and other "wickedly" complex problems defying quantitative solutions.

For example, educational researchers have for thirty years been trying to adequately quantify the most basic requirement in the field by researching what makes a quality teacher. Yet as the research supporting the No Child Left Behind legislation and the push for alternative certification pathways shows, there is no such data (at least none that can be agreed upon). The data and debate around this issue is legion (see, for example, Cochran-Smith 2001; Goldhaber and Brewer 1999; NCDTF 2004; NCTQ 2004). The basic point, though, is that if the educational field after all this time and research is still this stuck, woe to service-learning.

While such uncertainty appears to be absurd, it is also the end result and consequence of a quixotic search for absolute and quantifiable surety. None exists and any attempts to find it become quickly beholden to political pressures of which variables are measured and how. I do not deny that the quantitative move offers some basic guidance on some basic proxy variables. This is an important development. But to pin the legitimacy of service-learning on its quantification is to misunderstand how legitimacy ultimately works.

This is in fact the second reason why the quantitative move falters in the academy. Namely, the paradigms by which we see the world are inextricably linked to our value systems as legitimate scholars. Thomas Kuhn, in his classic *The Structure of Scientific Revolutions* (1967), posited that paradigms shift not because of rational discourse among objective scientists but because the old guard dies away to be replaced by the paradigm of the young turks. While the conservative status-quo nature of this view has been roundly critiqued, the underlying psychological framework seems sound (see, e.g., Gardner 2004): the more contested and revolutionary an issue, the stronger our resistance to it.

To again use an example from teacher education, a recent review of the literature on teacher change argued: "What we see expressed in these current studies of teacher education is the difficulty in changing the type of tacit beliefs and understandings that lie buried in a person's being" (Richardson and Placier 2001, 915). Thus after four years of coursework, field experiences, and self-selective dispositions toward becoming a good teacher, the vast majority of teacher candidates leave their programs believing pretty much what they came in with.

It is thus naive for service-learning advocates to believe that a large number of academics will be persuaded to accept service-learning simply because data shows it to have a statistically significant impact on any particular student outcome. As I have argued elsewhere (Butin 2005b), a simple thought experiment puts this lie to rest: if data showed that students' work with terminally ill AIDS patients negatively impacted student understanding of the social health system, would that be reason enough to stop the program? Probably not. Service-learning advocates would "question the validity and reliability of such data: How is 'understanding' being measured? Is success defined instrumentally (i.e., test grades) or holistically (i.e., emotional intelligence, long-term changes)? What was the timeframe of my assessment procedures? Did I use pre- and post-tests or interviews? Was there an adequate control group?" (102).

Of course if such data were consistent and long-term, there might be good reason to desist or substantially modify the service-learning component. But not only are most data not rigorous enough to warrant immediate acceptance, they function as only a small part in how we come to marshal evidence to support our views of the world. The quantitative move toward statistically significant measurement thus cannot, on its own, convince scholars to embrace or reject service-learning.

The third reason that the quantitative move in service-learning undermines rather than promotes the institutionalization of service-learning is exactly because it is quantitative. David Labaree (2004) has used Becher's typology of academic disciplines to nicely point out the decidedly problematic implications of a "soft" discipline (in this case, educational research) in search of a "hard" disguise:

> In order to create a solid ground for making hard claims about education, you can try to drain the swamp of human action and political purpose that makes this institution what it is, but the result is a science of something other than education as it is experienced by teachers and students. As I have argued elsewhere [Labaree 1997], such an effort may have more positive impact on the status of researchers (for whom hard science is the holy grail) than the quality of learning in schools, and it may lead us to reshape education in the image of our own hyper-rationalized and disembodied constructs rather than our visions of the good school. (75)

The scientific quantification of any human practice is what Max Weber (see Sica 2000) termed "rationalization." It is the attempt to order and systematize, for the sake of efficiency and (thus supposedly) progress, practices that were formally intuitive, haphazard, and grounded in heuristics rather than science. The point again is not that we should avoid scientific inquiry; it is that, simply put, this is not nor should be at the heart of service-learning. To promote service-learning in the academy through quantification is to buy into a paradigm not of its own making. The quantitative move may help service-learning scholars gain a certain legitimacy in the academy; what it will not do, though, is expand the boundaries of how to think about the academic. What it will not do is provide a decidedly different discourse of how service-learning should be institutionalized.

TOWARD A DIFFERENT KIND OF POLITICS FOR SERVICE-LEARNING

Thinking about service-learning as a form of politics has deep rhetorical resonance; it argues for practices and policies that are uplifting and transformational for all involved. Yet as the above sections have argued, such rhetorical resonance also has limited and limiting possibilities for institutionalizing service-learning in the academy. There are deep and specific pedagogical, political, and institutional barriers. Moreover, these limits are fundamentally linked to the undergirding theoretical presuppositions of contemporary service-learning theory and practice. I want to thus briefly explicate such presuppositions to begin to rethink and reframe how service-learning may be otherwise institutionalized, which I will do in-depth in the second part of the book.

Fundamentally, service-learning is presumed to be a politics by which to transform higher education. As such, service-learning becomes positioned within the binary of an "oppositional social movement" embedded within the "status-quo" academy. Moreover, this perspective reifies (and thus presumes) service-learning as a coherent and cohesive pedagogical strategy able to see its own blind spots as it pursues liberal and always liberatory agendas.

But such is not the case. The service-learning movement is an amalgam of, among other things, experiential education, action research, critical theory, progressive education, adult education, social justice education, constructivism, community-based research, multicultural education, and undergraduate research. It is viewed as a form of community service, as a pedagogical methodology, as a strategy for cultural competence and awareness, as a social justice orientation, and as a philosophical worldview. There is thus an immense diversity of oftentimes clashing perspectives cohabiting under the service-learning umbrella.

Likewise, the service-learning movement has oftentimes downplayed or glossed over the minimal social justice outcomes of service-learning practices. For all of the human, fiscal, and institutional resources devoted to service-learning across higher education, there are in fact very minimal on-the-ground changes in the academy, in local communities, or in society more generally.

I do not dispute that in isolated situations with unique circumstances profound changes have occurred. What I am simply pointing out is that service-learning should not have to bear the burden (nor the brunt) of being the social justice standard-bearer. To do so would be to set up an impossible causal linkage between service-learning and social betterment. Much scholarship, for example, can be marshaled to show that the divisions in our society based on categories of race, class, ethnicity, and language have in many cases become worse, not better; that democracy for all intents and purposes has become a spectator sport as most of us (and particularly youth) have disengaged from the public sphere; that the United States is the worst offender in the developed world of human principles and ethical norms for the treatment of its incarcerated population. Is this service-learning's fault? If service-learning succeeds as hoped in higher education and these conditions continue to deteriorate, does this mean that service-learning is to blame? The issues cited have much more to do with a host of interconnected economic, social, political, and legal policies than they do with the percentage of faculty implementing service-learning on any particular campus.

What all of this points to is that thinking about service-learning as a politics to transform higher education is a theoretical cul-de-sac. I do not doubt that service-learning may in fact become deeply embedded within higher education. Yet I suggest that if it does not account for the pedagogical, political, and institutional limits enumerated, it will do so without any grounding for its long-term sustenance. It will do so by giving up any analytic opportunity to understand how and why it is ultimately deeply limited.

All of the theoretical presumptions of the service-learning movement just enumerated position it as a gleaming grand narrative. Service-learning scholars and activists want service-learning to be all things to all people. Service-learning wants to roam free across disciplines, across institutions, across society. It wants to change and transform any and all obstacles in its path. It wants freedom.

But that is not how things work in academia. Higher education is a disciplining mechanism, in all senses of the term. And that is a good thing. For to be disciplined is to carefully, systematically, and in a sustained fashion investigate whatever it is one is interested in doing, be it building bridges, changing communities, or understanding Kant. If the service-learning vision turns into a mirage—as scholars begin to carefully examine its claims and outcomes—the grand narrative will implode. For there is no mechanism by which a grand narrative can prevent itself from being questioned and critiqued once it has become a part of the academy. That is the basis of higher education and that is where, for better or worse, service-learning wants to be positioned.

The possibilities for service-learning, I will thus suggest in later chapters, lie in embracing rather than rejecting the very academy the service-learning movement is attempting to transform. More precisely, rather than continue to think about service-learning as a politics to transform higher education and society, we might more fruitfully reverse the terminology to use all of the tools and structures of the academy to analyze and reshape what we may mean and do in service-learning. This is to speak about service-learning as akin to an academic discipline: having the ability to control its knowledge production functions by internally debating and determining what issues are worthy of study, by what modes of inquiry, and to what ends.

This, in turn, presumes a plurality of perspectives of what service-learning is and should be. It presumes that the scholarship surrounding service-learning is not solely centripetal or convergent in focus. Service-learning may no longer claim that it will change the face of higher education. But it may allow for the careful and systematic examination of issues of voice, community engagement, reciprocity, etc. This is not radical and transformational change. This is disciplined change. It is the slow accretion, one arduous and deliberate step at a time, of contesting one worldview with another. Some of it is blatantly political. Some of it is deeply technical. Much of it is debatable, questionable, and modifiable. Just like any good academic enterprise. And it is this that is truly transformational.

At present, though, such heteroglossic analysis and critique are largely absent in the service-learning field. If service-learning is presumed to be "simply" a universal, coherent, and neutral pedagogical practice, then such an absence is understandable. But such is not the case. As I demonstrate in the next chapter, service-learning can in fact never be transparent nor simply or easily aligned solely with what it is attempting to teach. It thus becomes incumbent for scholars committed generally to a scholarship of engagement and specifically to service-learning to probe the limits of service-learning in higher education. For without an explicit articulation of its own limits, service-learning may be doomed to a limited and limiting model of transformation.

CHAPTER 3

THE POSSIBILITIES OF SERVICE-LEARNING

The previous chapter articulated the pedagogical, political, and institutional limits of service-learning in higher education. In this chapter I want to take a final step in the analysis by suggesting that in fact all modes of service-learning are limited, or more precisely, all modes of service-learning are self-undermining. Thus all modes of service-learning not only have external limits; they have internal limits as well. And this, I want to argue, is the most powerful and transformative possibility of service-learning in higher education.

Such an argument may at first appear not just counterintuitive but plain absurd. Yet the key to unraveling and dismissing this absurdity is to first realize and acknowledge that all modes of pedagogy are self-undermining. In fact, most modes of pedagogy are so inherently and fundamentally flawed as to be inoperable. This is not a grandiose claim; it is just a rephrasing of the common wisdom and long-standing research that traditional didactic, lecture-based instruction is the worst form of instruction for the vast majority of our students and for most types of content (Angelo and Cross 1993; Bligh 2000). Put simply, lecturing to a roomful of eighty students has so many internal constraints and self-undermining enactments (such as the inability to gauge whether students understood what you just lectured about) that it is common practice to talk about "teaching to the top twenty percent," irrespective of whether one says this in defeat or in shocked incredulity.

Once it is acknowledged that all forms of pedagogy have their own internal constraints and subversions and that the (strategic) key to good teaching as such is to align instruction, curriculum, and context, it becomes possible to rethink and reframe the power and potential of service-learning. Specifically, it becomes possible to view the service-learning experience as a nontransparent activity that necessitates a constant undercutting of and attention to the academic content being taught. This, in turn, necessitates the exposure of implicit presumptions and power dynamics within service-learning and content knowledge; it fosters deep, consequential, and long-term experiences within the field; and it fosters an openness to others' voices and perspectives, especially of those within communities partnering with the educational institutions. As I detail in the next section, this allows us to make visible and begin to dismantle what I term the myth of the stable educational experience, the myth of the singular community, and the myth of an agreed-upon justice.

The implications of this for higher education policies, practices, and structures will be detailed in the second part of this book. What is key for the moment is to suggest that service-learning—when properly understood—is an ideal pedagogy for transformation in that it allows (in its antifoundational form) students and faculty an opening and window into its own constraints and subversions. Put otherwise, service-learning—by its very nature of being a community-based, experiential, and embodied experience that is culturally saturated, socially consequential, politically contested, and existentially defining—makes visible the complexity of the content and context of the educational experience.

The transformational potential of service-learning in higher education thus rests in its academic capacity—what I term (and expand upon in the concluding section of this chapter) as "justice in doubt." For service-learning frees us from the false notion of controllable teaching of controlled subject matter, from knowledge as static, and from truth as fixed. This is justice-oriented education in that it allows a hyperengaged and community-based pedagogy that attends to the nontransparent aspect of service-learning under which there is never an educational experience without remainder. By this I mean (as I describe in detail in the next section) that service-learning is never a transparent activity that accomplishes exactly what I as the instructor want it to accomplish. There is always a slippage exactly because of the reality that service-learning is an embodied and experiential activity that cannot be contained—as much teaching and learning attempts to be—within the four walls of the classroom and the covers of the textbook. There are always "remainders" in the experiential experience that subvert our attempted practices.

Such "remainders," as such, pull us up short in our seeming attempts to simply and consistently and with statistical significance move always toward greater cultural competence or equity. I acknowledge that these are deeply important goals for our students, colleges, and communities. Nevertheless, as I will detail, such seemingly clear-cut perspectives actually obscure and thus all too often obfuscate the immensely complex and contested issues of community engagement in our pluralistic democracy. Put otherwise, and as I have shown in the previous chapter, the seemingly direct "strong overcoming" of students' lack of content knowledge, cultural competence, or sense of social justice may in fact face much greater barriers and a lower "ceiling" for success.

It may thus be necessary to speak more of a "weak overcoming" (Butin 2002) that acknowledges the always inherent tensions and slippages of our service-learning practices to move toward a more realistic, and ultimately, more justice-centered vision and practice of our community-based engagements. This may be deflationary for grand claims, but it is empowering for sustained theories and practices that can truly work. As such, this chapter argues that most traditional modes of service-learning are inadequate for truly fulfilling their goals. I thus document how most modes of service-learning are self-undermining in relation to their explicit goals as they do not fulfill the theoretical or pedagogical conditions they themselves have set. Specifically, I demonstrate that each mode of service-learning already has within itself the condition of its own subversion.

Several implications arise from this explication: first, that different modes of service-learning have distinctly different pedagogical and theoretical limits; second, that such internal self-subversions function as the upper limit cases (the "ceiling") for deeply embedding service-learning in higher education; third, that the hoped-for fulfillment of liberal- and justice-oriented claims may be overreaching and out of reach for most service-learning models; fourth, and finally, that through this rearticulation it becomes possible to rethink the potential of service-learning as well as to recreate the means by which higher education deeply institutionalizes service-learning within the academy.

I begin by explicating how each mode of service-learning—the technical, cultural, and political—is undermined at both the level of its performance and at its very condition of possibility. I then suggest that such self-undermining must be—if not embraced—at least acknowledged and worked through in service-learning practice. I provide two specific examples of service-learning practices that demonstrate and enact such a vision of service-learning. I then conclude this chapter by focusing on the implications of such practices for fostering deeper engagement and awareness of a community's voice.

The Limit Cases, Performative Contradictions, and Internal Subversions of Service-Learning

I use my previously articulated conceptualizations of service-learning—technical, cultural, and political—as the lens through which to examine each mode's self-subversion and limit. I do so because traditional models (e.g., Bringle and Hatcher 1995; Furco 1996; MJCSL 2001; NCSL 2002; Hayes and Cuban 1997; O'Grady 2000; Westheimer and Kahne 2004) have multiple theoretical problems that cannot be supported—normative teleology, ethical foundationalism, a stage theory model of individual progress from charity to justice—and obscure more than they reveal, particularly in regard to how faculty actually think about, make use of, and connote value judgments for service-learning.

As I discussed earlier, a *technical* conceptualization of service-learning is focused on its pedagogical effectiveness whereby it is one amongst many pedagogical strategies that serve the function of better teaching for better learning: for example, an excellent way to teach about the impact of poverty on families is to work with actual families in actual poverty within the context of an academic course that uses multiple other texts, reflections, and assignments related to the topic.

A *cultural* conceptualization is focused on the meanings of the practice for the individuals and institutions involved so that, for example, service-learning may be seen as a means to help students increase their tolerance and respect for diversity and for academic institutions to promote engaged citizenship. It is, by the way, this differentiation (and tension) between the civic and the academic that is at the heart of contemporary debates about the role that service-learning can play in the mission of higher education. Advocates suggest an expansive view that embraces community engagement and the civic-minded implications thereof (see, e.g., Benson et al. 2007; Bok 2005; Colby et al. 2003), and critics argue that the academy should attend to its primary mission of knowledge production and dissemination and not dabble in volunteerism (Fish 2008; Neidorf 2005).

Finally, a *political* conceptualization is focused on the promotion and empowerment of the voices and practices of historically disempowered and nondominant groups in society. This perspective animates and informs the service-learning movement; it is also, though, the most contentious given recent controversies over the need for "balance" in the curriculum and the seeming lack of "academic diversity" in higher education due to the liberal and radical leanings of the professoriate. It is here that the distinction between cultural and political perspectives is brought into sharpest relief: civic engagement (e.g., voting, volunteering) is made of eminently cultural practices that are embraced by higher education leaders and funded by federal grants; activism and "social justice," though, are deeply contested political notions that incite litigation and censure by these very same constituencies (see, e.g., ACTA, 2006).

It should be noted that any particular service-learning practice can have aspects of all three types. The semester-long tutoring of underperforming high school students in math can help college students understand how youth make systematic conceptual errors (a technical perspective); students can gain insight into working with youth of a different socioeconomic status and/or ethnic background (a cultural perspective); and they can explicitly link the tutoring with college preparation to support college admission of such underrepresented youth (a political perspective). Additionally, as will be highlighted in depth later on, all three modes are of the same "kind" in that they all view service-learning as a type of intervention that facilitates a better process toward reaching a specific predetermined goal, be it comprehension of math, cultural competence, or more equitable access to scarce resources.

A final note is that my analysis of the internal subversions within service-learning does not examine the antifoundational perspective outlined in the first chapter. This is because—to put it quickly and simply for now—antifoundational service-learning is not encumbered by the myth of its own transparency. This does not mean that antifoundational service-learning is free of internal subversions and limits. It is just that it is the only one of the four outlined perspectives that acknowledges and works through, rather than against, such subversions and limits. This distinction will become clear at the conclusion of this chapter, and will be expanded upon in the second part of this book in general and in Chapter 7 in specific.

The rest of this chapter traces how each mode of service-learning bears its own subversion and limits. Figure 3.1 provides an overview. The key to note is that each mode of service-learning has an explicit goal—be it conveying content, fostering diversity, or enhancing equity and tolerance—that is undercut by the very nature of the service-learning

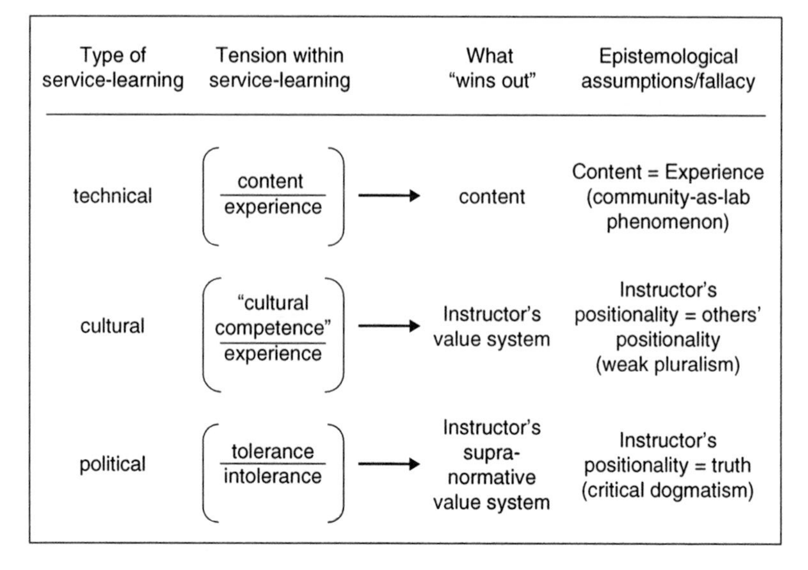

Figure 3.1 The limits and subversions of service-learning practices

process as a community-based, experiential, and embodied experience. In each case the service-learning mode is beholden to an internal fallacy that functions as the limiting factor for the potential and power of the service-learning experience.

Thus, to briefly summarize the first situation, an instructor working from a technical perspective presumes that the experience within service-learning is equivalent to and in support of the content knowledge being taught. This is what I term the myth of the stable experience; that is to say, what students do in the field always and without remainder maps onto the course's content knowledge. Yet "content" and "experience" operate within an implicit and hierarchical binary, with content knowledge "higher than" and thus winning out over the experience. Such an occurrence operates within what has come to be viewed as the standard problematic of "community-as-lab": institutions of higher education simply entering and leaving communities to suit themselves and their goals and prerogatives. As I will show below, each mode of service-learning practice has its own tensions that ultimately subvert any simple attempts to use service-learning as a transparent practice without remainder.

Before moving forward to deeply explicate each type of service-learning, it is necessary to heuristically differentiate between the performance of service-learning and the condition of possibility that makes particular performances possible. Service-learning has been embraced in higher education exactly because it is a community-based, experiential, and embodied activity. It is seen as a pedagogical method by which to move beyond the walls of the classroom and the covers of the textbook to incite and excite students about the deep complexities and controversies of our world. This is its condition of possibility. Such a condition of possibility, in turn, has been taken up (i.e., performed) in different ways and with different goals. What I show in this section is that each model of service-learning has distinctive limits; each goal is undone since the performance and/or the condition of possibility of such a performance carry within themselves the subversion of the attempted performance.

The theoretical grounding of this argument comes from Derrida's (1976) notion of *différance*. My interest here is not specifically in Derridean notions of meaning or postmodern interpretations of teaching (see, for this, Biesta 1998, 2006; Biesta and Burbulles 2003; Bingham 2008). Rather, it is to simply take up a key insight of antifoundational work that the "content" (be that the text, the teacher, or the field-based experience) can never truly and totally "erase" itself to allow the content to speak for itself. In fact, the very attempted act of erasure (e.g., believing the text or teacher is neutral and/or transparent) propels the undercutting and subversion of the attempted goal.

In a technical perspective, the performance of service-learning is undermined by its inherent condition of fostering engagement with real-world issues in real time. This is, again, the myth of the stable experience in service-learning. For example, an instructor in a mathematics course may assign students to engage in math tutoring at a local elementary school to demonstrate the conceptual errors that youth make at different developmental stages. It is thus hoped that the service-learning experience will be aligned with (without remainder) the course content. Yet all experiences are polyvocal. The undergraduate students may begin to ask why girls seem to be less motivated than boys and why some youth (and perhaps even teachers) perpetuate stereotypes of who is good in math and who is not, or students may want to make connections to their own math phobias as youngsters. Each of these scenarios demonstrates a possible "remainder" that is present above and beyond the supposed content knowledge being taught.

Each of these potential issues may of course be inherently interesting and academically important, but just not for this particular course. Instructors must then make difficult decisions as to whether to delve into academic fields (e.g., gender bias in the classroom) that may be far from their expertise, to ignore or minimize such discussions and risk disenfranchising interested students, or to develop entirely new readings that cut into the already planned curriculum. When any of these scenarios occur, the instructor's initial desire for better teaching of content knowledge will trump the (seemingly errant) experiences brought into class. After all, faculty who bought into a technical model of service-learning did so exactly because they wanted a more effective pedagogical methodology for conveying their content matter. As soon as the particular pedagogical methodology is no longer effective (since it takes up precious classroom time to discuss math phobias or cultural stereotypes concerning girls' success in math), it is of no use for the faculty's end goal. This is just as true of PowerPoint and project-based learning as it is of service-learning. When we as teachers find certain teaching strategies not working, we stop using them.

Flowers and Temple (2009) offer a vivid account of such a dilemma—they had attempted to prepare their undergraduate students to tutor through an America Reads program in a literacy course. The key tension, they suggest, was the "extensive need to train students to be effective tutors . . . how much time should the students spend in class learning to be tutors? The course on literacy is a broadly based liberal arts course and not a training program" (91). The problem, of course, is that neglecting the "training" component can seriously shortchange the very reason for doing the service-learning in the first place: helping the youth being tutored. "In order for the literacy course to serve the America Reads program well—and also to be prepared for tutoring the students in class—we have had to devote much class time to teaching and assessment matters, and we have devoted less time to other topics" (102). At some point, as course content gets shortchanged, faculty may say, much like Flowers and Temple (2009), "But there is never enough time" (102), and as such move away from their original intent of offering such a service-learning experience.

The reason for the implosion of the performance of service-learning from a technical perspective—where experience is trumped by content knowledge—is also in this case at the heart of the limit of its condition of possibility. For in the technical perspective, the instructor must implicitly conflate the service-learning experience with the hoped-for course content. An instructor may believe that teaching math to elementary school students will help her undergraduates better understand how youth make conceptual errors at different developmental stages. Experience, from this perspective, equals content. Yet this conflation is actually predicated on the construction of and reliance upon an implicit and hierarchical binary made of the course content and the service-learning experience.

The experience is implicitly understood to be in the service of and inferior to the overarching goal of content knowledge acquisition. When the means (of service-learning) undercut the goal (of content knowledge acquisition), the binary is revealed. Put otherwise, a technical perspective—by its very nature of focusing on particular content knowledge—is inherently open to being undermined at the level of performance and in the very condition of possibility by the culturally saturated service-learning experience. The instructor may, of course, take up the culturally saturated issues raised by the service-learning experience, but then she is no longer teaching from a technical perspective, and, as such, must attend to the issues raised by whichever other perspective she takes up.

Instructors working from a cultural perspective face an analogous dilemma of inherent "cracks" both in the empirical performance of the service-learning experience and in its conditions of possibility. On the empirical performative level, the very experience which, it is hoped, will foster an openness to the diversity and plurality of local and global communities may in fact reinforce students' deficit perspectives on the other. Students volunteering in a homeless shelter may see their worst stereotypes reinforced by violent, sexist, or demeaning behavior. They may now have "data," however anecdotal, that supports their predetermined convictions. This is what I term the myth of the singular community.

The hoped-for openness and understanding envisioned by a cultural model of service-learning thus stops at the limits of its own openness and understanding. Specifically, the experiences in the community may be antithetical to the knowledge, skills, and dispositions we as faculty are attempting to foster: when *we* believe in letting children speak and the family being visited believes that children should be seen and not heard; when *we* believe that women have control over their own bodies and the community organization one is volunteering in tells callers that only God has the right to kill a fetus; when *we* believe that skin color is an irrelevant basis by which to judge others and the youth one is tutoring continue to refer to each other in what we consider to be highly derogatory terms. Such culturally saturated experiences force faculty to confront the limits of their control over what students are exposed to, and, as such, over what can and cannot be taught through such a service-learning experience.

Stoecker and Tryon (2009) found such situations all too common when they interviewed community organizations about working with service-learning courses at the University of Wisconsin at Madison. Stoecker and Tryon (2009) comment that "community organizations, of course, do not expect the people they serve to be an expandable source of student enlightenment," and then quote a community leader stating that "people have to be able to come here [the service organization] and expect nondiscriminatory service, whether they are a white supremacist or a lesbian couple. So people who can't show professional objectivity can't volunteer here" (123).

This is a classic case of weak pluralism, or what Stanley Fish (1999) terms "boutique multiculturalism"; he characterizes the phenomenon and its self-corrosion as follows:

> Boutique multiculturalism is characterized by its superficial or cosmetic relationship to the objects of its affection. Boutique multiculturalists admire or appreciate or enjoy or sympathize with or (at the very least) "recognize the legitimacy of" the traditions of cultures other than their own; but boutique multiculturalists will always stop short of approving other cultures at a point where some value at their center generates an act that offends against canons of civilized decency as they have been either declared or assumed. (56)

A cultural model of service-learning thus only functions by erasing its very own positionality until it is caught up short by encountering an experiential situation that forces it to acknowledge its own limits. This is the implosion encountered at the level of the condition of possibility. As Fish (1999) continues, the boutique multiculturalist "does not and cannot take seriously the core values of the cultures he tolerates. The reason he cannot is that he does not see those values as truly 'core' but as overlays on a substratum of essential humanity" (57). Tolerance to other perspectives functions so long as the encountered perspectives mirror one's own internal (and hidden) norms. As soon as this boundary is crossed, the potential for a cultural perspective of service-learning collapses.

Thus we may *want* to believe that children should have a chance to articulate their perspectives and in general their voices be valued and heard. But this becomes more than problematic when the host family disagrees, or more strongly, complains to the volunteer coordinator of a college student's interference in family matters. It becomes an undermining of the very goals of openness that the instructor was attempting to foster. In the face of such resistance (implicit and/or explicit) from the community, it becomes impossible for the instructor to continue positioning such goals and ideals as common or universal.

Thus whereas the technical perspective functioned through an implicit content/experience binary, the cultural perspective functions through an implicit and hierarchical binary between the instructor's and the community's key principles in their respective value systems. Thus the instructor who believes her goals (of implicitly teaching a particular value system) are undermined (by the divergent value systems encountered in the service-learning experience) will of necessity stop using the means she had originally been using. This process of subversion, it should be noted, functions exactly in the same manner for both the technical and the cultural mode of service-learning. And, again, it is worth noting that if the instructor is in fact prodded by the community's value system to rethink her own principles and thus reposition her course goals, she will no longer be doing cultural service-learning but something different.

Finally, a political perspective is undone at both the performative level and in its condition of possibility by a "critical dogmatism" that leaves unquestioned its own foundational underpinning that discounts alternative perspectives. This is what I term the myth of an agreed-upon justice. On a performative level, such a political perspective is undermined by its explicit embrace of a distinctively partisan orientation. The linkage of service-learning to social justice inherently presumes a dichotomous liberal/conservative spectrum with service-learning meant to function as a mechanism to move individuals from the (political) right to the (social justice) left. This is traditionally described as helping students move from individualistic to structural understandings of societal problems, and from passive acceptance to collective action.

Yet the very explicitness of such an agenda incites responses such as the ones I outlined in the previous chapter (e.g., David Horowitz's "academic bill of rights") in that students have just as much "academic freedom" to learn alternative perspectives as faculty have to teach a singular "truth." This counterdemand for "balance" derives exactly because conservatives believe that college students are only being exposed to one side of the academic story and, as such, should be allowed to be taught "the controversy" rather than be "indoctrinated" by distinctly liberal and radical perspectives. The performance of service-learning from a political perspective thus creates the condition for its own undoing, since it radicalizes dialogue as between extremes.

The performance of service-learning from a political perspective, it should also be pointed out, is pedagogically suspect. As Dewey (1938) once whimsically noted, "frontal attacks are even more wasteful in learning than in war" (176). By attempting to directly "deposit" the "correct" knowledge in students, a political perspective risks undermining its own goals by fostering student resistance to the process of being educated.

A political perspective of service-learning is undermined at the level of its condition of possibility due to two implicit and nested binaries. The first binary is analogous to and a stronger version of boutique multiculturalism. Namely, a political perspective of service-learning is what Fish would term a "strong multiculturalist" one because it takes as its first principle not simply a taken-for-granted "humanism," but the very core of such a presumption: tolerance. Thus, for example, the American Civil Liberties Union (ACLU) will defend the free speech rights of organizations (such as the Ku Klux Klan) that it may vehemently disagree with and even abhor. Yet, Fish argues, such tolerance and respect for diversity always meets a limit case whereby it cannot accept the founding conditions of the other. As Fish points out:

> The trouble with stipulating tolerance as your first principle, however, is that you cannot possibly be faithful to it because sooner or later the culture whose values you are tolerating will reveal itself to be intolerant at that same core . . . Confronted with a demand that it [the "tolerant" culture] surrender its viewpoint or enlarge it to include the practices of its natural enemies—other religions, other races, other genders, other classes—a beleaguered culture will fight back with everything from discriminatory legislation to violence. (60–61)

This is a lose-lose situation for the strong multiculturalist, for she either has to be tolerant and embrace a culture that is intolerant and as such, by implication, be intolerant to the others who are being oppressed by the intolerant culture; or she must condemn, from some supranormative position, the intolerant culture that is in fact being intolerant and show that this will not be accepted. In either case, Fish points out, "it turns out that strong multiculturalism is not a distinct position but a somewhat deeper instance of the shallow category of boutique multiculturalism" (61).

The only differentiation, and this is one of degree rather than kind, is that the strong multiculturalist makes explicit her critical dogmatism. And this is the second binary. For the strong multiculturalist who engages in service-learning from a political perspective must by necessity always appeal to a supranormative position that is deemed both beyond critique and inherently superior to all other positions that may challenge such a positioning. When a faculty member assigns a community-based project that is deemed to support the goals of social justice, she, by necessity, must believe that her vision of social justice is accurate; else, why have her students engage in such practices?

Nonetheless, as Biesta (1998) demonstrates, such a stance of strong multiculturalism that engages in critical dogmatism—a criticism that functions by a criterion (e.g., "social justice," "emancipation") that is itself "kept out of reach of the critical operation" (7)—is by its very definition subverted by itself. For critical dogmatism must either succumb to an "infinite regress" whereby its foundational conditions must always be questionable, a "logical circle" in that all justifications for critique or rationale for specific criteria are always already in need of being justified, or a "breaking off the attempt [of justification] at a particular point by *dogmatically* installing a foundation" (8).

If I believe in "tolerance" as my first principle (or "justice" or any other so-called neutral principle that Fish showed as functioning as an "empty vessel"), I must either consistently (and infinitely and regressively) find undergirding foundations that, through questioning and critique, can only be supported by ever deeper foundations for tolerance; or by referring to other seeming first principles (e.g., "justice," "human rights") that spiral back onto themselves; or by simply and finally—caught in such regressive loops—stating with a finality that allows no exterior questioning that "tolerance" must serve as the "true" foundation. The very idea of service-learning for social justice, in other words, collapses under its own presumptions.

So when I as a faculty member am attempting to teach social justice, and the very way I do so instead leads to demands for "balance" from external constituencies, incites resistance by my own students, and is confounded by others out in the community who have their own first principles of (what I deem as) intolerance, I must retreat and find new ways by which to accomplish my goals. I thus drop service-learning. So, again, the very means of attempting to get to a predetermined goal confound the attempt, thus forcing the abandonment of allegiance to service-learning as the pedagogical "delivery mechanism" of choice. And, again, if the instructor is somehow, through the service-learning experience, convinced that she is no longer superior, she is also no longer doing political service-learning.

It should by now be (painfully) obvious that all three modes of service-learning—technical, cultural, and political—are undermined at both the level of performance and in the very condition of possibility of the performance. Moreover, such undoing functions in the exact same fashion for all three modes. Namely, what was originally considered an ideal methodology turns out to be not just idealized, but unreachable. The epistemic certainty with which service-learning was engaged and that stood at the heart of the decision—that content and experience were aligned (from a technical perspective); that faculty and community principles were congruent (from a cultural perspective); that first principles were not only the same but mirrored the faculty's own foundational presumptions (from a political perspective)—turns out to be the very place where the subversion begins.

Service-learning, in other words, is subverted from the inside out. And it is here, I suggest, that we must begin to look for ways to engage service-learning with fewer opportunities for being beguiled by the illusion that service-learning is an answer; it is instead, as I have noted, a "weak overcoming."

TURNING SERVICE-LEARNING INSIDE OUT

It is worth repeating that there is no unadulterated and "pure" form of service-learning free from the limits and subversions I have just explicated. Indeed, there are no other ways of "doing" service-learning except through the technical, cultural, and political perspectives of service-learning that I have examined. But I suggest it is possible to begin at exactly the conditions of possibilities that are otherwise ignored. By beginning with the realization that service-learning—in its very performance as a community-based, experiential, and embodied experience—is disturbing, it becomes possible to work with this process of undoing rather than simply be undone by it. This process, it turns out, has a somewhat off-centering function in that what occurs is not just simply a technical, cultural, or political end result. But I am getting ahead of myself. Let me offer two performances as an example.

The first example comes from the field of criminal justice. Lori Pompa (2005) developed the Inside-Out Prison Exchange Program over a decade ago at Temple University, Philadelphia. Pompa brings together undergraduates (the "outside" students) and incarcerated men (the "inside" students) at a maximum security prison within the context of a semester-long academic course. Both groups work together on texts concerning, for example, the criminal justice system, deviancy, restorative justice, and ethics. This working *with*, rather than working *for*, the incarcerated men offers undergraduates authentic and intentional pedagogical encounters that force students to make explicit their assumptions on prisons, crime, and punishment and analyze them. It also offers the incarcerated men the opportunity to further their education and begin to better understand their particular situations within a larger theoretical context.

Pompa (2005) deliberately modified her initial course from monthly visits to the prison to a fully engaging experience: "Having class inside a prison is compelling—an experience that's hard to shake. And that is one reason we do it. I don't want my students to shake these encounters easily; in fact, I want the students to be shaken *by* them" (302). Central to the experience is that the inside and outside students, once together, encounter each other as equal dialogue partners concerning the specific issues and texts of the day. This type of experience is highly disturbing to undergraduates used to either passively absorbing knowledge or, in the case of traditional social activism strategies, being the "givers" or "servers." As Pompa states, "In taking class together as equals, borders disintegrate and barriers recede. What emerges is the possibility of considering the subject matter from a new context—that of those living within that context" (305).

A second example comes from political science. Susan Dicklitch (2005) developed a course at Franklin and Marshall College entitled *Human Rights-Human Wrongs* that explored issues of human rights in general and the United States' asylum policies in particular. Students served as researchers for community partner organizations on asylum seekers' cases at York County Prison, Pennsylvania, the second largest detention center in the United States for asylum cases to be decided by the Bureau of Citizenship and Immigration Services and the Department of Homeland Security. Through interviews with the detainees (who come primarily from sub-Saharan Africa and Central America) and intensive research on immigration policies, human rights theory, case law, and the specific situations of each asylum seeker's story and country of origin, students created culminating immigration-court-ready documents and legal briefs for the detainees.

Since most detainees could not afford a lawyer (nor were they entitled to the equivalent of a public defender), had limited English proficiency, and were completely unfamiliar with U.S. immigration laws, Dicklitch's students were oftentimes detainees' sole link to any form of legal representation. It is important to note that students were not attempting to simply free these detainees; some detainees had falsified their background and others' situations did not meet the federal requirements for asylum. What is of relevance is the nature of the learning:

> Students did not have to be pushed and prodded to do their work. This is because this course had more than just the traditional course pressures i.e., getting a good grade, being able to answer professor's questions, not letting down the community partner. The students knew that if they did not put in the time to properly research their asylum seeker's story, case law, and find evidence in human rights reports to substantiate the claims, their asylum seeker would most likely get deported. And, if in fact the asylum seeker was telling the truth about his/her human rights abuse, deportation could mean further torture, abuse, or even death. Other human beings, from different cultures, speaking different languages, living completely different realities were depending on my students to make sure that their story got heard. (132)

The impact on students and on the asylum seekers has been profound. To date, six out of the twenty-eight cases have been granted asylum, the others have either been denied or are still under appeal (these are impressive numbers given that the national asylum grant rate is fewer than one in five). Most vividly, one of the detainees who was granted asylum came and spoke to Dicklitch's class upon being released: "Without the student's help, I would not be standing in front of you now, free, telling you my story" (37n10).

These examples, I suggest, demonstrate a type of service-learning that embraces and works with the internal subversions of its own practices. The situations students encountered necessitated reflection, analysis, and action exactly because students had to work within and through the experience's attempted subversions. In fact, the service-learning experience was the subversion; walking into a prison or listening to a detainee's story functioned as both the content of the course and the disruption to the students and the course that itself (the disruption) functioned as the content. Thus rather than attempting to construct a transparent service-learning experience without remainder, both of these courses embraced the "remainders" of the course—i.e., being shaken up by the slamming prison gates, being lied to by the detainee—as an integral component of understanding the issues and content at stake within the academic examination of the issues under analysis.

For Pompa, outside students were shaken up such that they had to ask fundamental questions (e.g., "who are prisoners?" "what is crime?" "when is justice served?" "what is freedom?") and at the same time engage with the inside students sitting side-by-side with them to discuss, for example, Sartre, and learn how to "do time" rather than time "doing them." For Dicklitch, students had to parse and weigh detainees' stories within (at minimum) the context of international law, local customs of the detainees' home countries, the internal consistency and integrity of the story told, and potential cross-cultural miscommunication, all the while aware that this was a real-life situation playing out in real time with real consequences. Not only did Dicklitch's undergraduates have to internally struggle with the reality that asylum seekers were explaining why death may be the end result of deportation, while they themselves were soon to return to their dorm rooms, such incongruities themselves had to be explicitly grappled within the class vis-à-vis issues of ethics, procedural justice, and immigration law.

The point here is that no type of service-learning—technical, cultural, or political—can function without remainder; that is, no service-learning course can stay "true" to its seemingly direct and transparent goals. If Pompa had wanted, from a technical perspective, to better teach about prison conditions, prison rights, and criminal justice, the multiplicity of seemingly tangential issues that cropped up in a prison visit or encounter might have overwhelmed her predetermined and focused goals for covering particular content knowledge. In fact, no matter how wide-ranging and broad her syllabus may have been, it would have been impossible to presume and prepare for all of the contingencies and situations encountered within the prison experience. Thus Pompa could have attempted to ignore and avoid such "tangential" issues; she could have embraced such "teachable moments"; or she could have cut back on such experiences to focus her class on the relevant coursework issues. But in all of these scenarios, the technical focus of the service-learning would have been undermined.

Likewise, if Pompa or Dicklitch had wanted to primarily teach through cultural or political modes of service-learning about, respectively, openness to difference or to advocate for the human rights of the oppressed (be it of prisoners or detainees), the goals would have been subverted and undermined as well. Pompa's outside students could not ignore or avoid that they were sitting next to individuals incarcerated for crimes such as murder and rape. Dicklitch's students could not avoid or ignore that some detainees were lying about their situations to remain in the United States. And neither Pompa nor Dicklitch could control or prevent prisoners and detainees from making provocative and intolerant statements or holding views antithetical to themselves or undergraduates. If allowed to function on their own, such statements and first principles would have undermined the instructors' goals. And, again, the instructor would have either had to retreat from the content, the experience, or her own principles and goals. The service-learning experience would not have functioned as planned.

These two examples of service-learning practices may be extreme in their depth and complexity. But they reveal that service-learning is never transparent nor without remainder. The myth of transparency is in fact what ultimately serves to undermine and disrupt the service-learning experience. From a technical perspective it is the seeming transparency of the experience as fulfilling the goals of content acquisition that undermines the experience. From a cultural perspective it is the seeming transparency of the experience of the community's positionality as mirroring the instructor's and course's cultural and ethical positionality. From a political perspective it is the seeming transparency of the notion of social justice as supranormative and mirroring, again, the instructor's and course's cultural and ethical positionality.

But, as I have shown, it is not possible to have a pure mode of service-learning that does not "slip" from its goals given the context within which it is being played out. And neither Pompa nor Dicklitch presume that one can do service-learning without slippage. Or more precisely, both Pompa and Dicklitch employ pedagogical strategies through the service-learning experience that presumes that nothing within the academic course is self-evident or without remainder. All components must be examined, questioned, and reintegrated into the academic content and experiential component. (I should note that Pompa and Dicklitch may in fact have a different hermeneutics to explain what they do [see, e.g., Pompa, 2002]; this, though, does not obviate my own hermeneutical appropriation and explication of their strategic moves.)

There are multiple implications of such a type of service-learning that become operationalized in important ways in the fostering of students and faculty buying into it, in the engagement of community voices, and the structuring and scaffolding of service-learning experiences. I explore these in depth in Chapter 7. For now it is adequate to conclude with the implication that it is only through doubt that justice becomes appropriate. Understanding how service-learning operates through and within subversions allows us to understand the potential of service-learning as justice-oriented education and what I term as "justice in doubt."

Toward a Conclusion on the Value of Justice in Doubt

I have attempted to show that traditional models of service-learning are fundamentally limited. Such limits, moreover, reside in the very essence of their respective performances and serve to undercut the hoped-for goals initially articulated by each model of service-learning. Service-learning is disturbing its own goals and aspirations at the very level of its condition of possibility and performance. More specifically, all traditional models are operating within the context of an epistemic certainty that presumes (wrongly, it turns out) a particular intervention is the best means toward a particular and predefined goal.

Yet this chapter has suggested that there is no such thing as the surety and transparency of the service-learning experience. What Pompa and Dicklitch constructed is an antifoundational service-learning that works within and through an epistemic doubt that must constantly examine and take account of its own foundations and operations. To put it simply, antifoundational service-learning operates from the presumption of service-learning-as-question rather than service-learning-as-answer. To believe in the transparency of service-learning is to believe that in engaging in the service-learning experience is to gain a particular answer in a particular way. What I hope I have shown instead is that it is possible to foster truths and answers, but only as by-products of the service-learning experience that works by questioning answers rather than answering questions.

This is "justice in doubt." By this I do not just mean the seeming play of words that our goals of justice-oriented education may never be reached. This is in fact the case in that we can never truly control what students leave our courses believing. And this is what Fish (2003) meant when he wrote about "aiming low" in our goals and expectations of what can and cannot be taught in the college classroom (and which I take up in detail in Chapter 7). But more specifically, I mean that one cannot approach complex and contested notions that are culturally saturated, politically volatile, and existentially defining without a constant and careful vigilance regarding the falsehoods of surety and the transparency of the (so-called) truth.

A final analogy may be applicable as it relates to Pompa's *Inside-Out* program. The fostering of doubt through Pompa's service-learning experience in the prisons is meant to better understand the complexities of the criminal justice system. It is not meant to trivialize or foster a radical relativism about crime, punishment, or justice. Pompa does not expect nor suggest that her undergraduates leave her course believing that all "lifers" should be immediately pardoned. Antifoundational service-learning is rather about dismantling implicit and thus oppressive binaries.

In 1971 Foucault joined several other French intellectuals to form the Prison Information Group. Its goal, Foucault argued, was not prison reform:

> The ultimate goal of its interventions was not to extend the visiting rights of prisoners to thirty minutes or to procure flush toilets for the cells, but to question the social and moral distinction between the innocent and the guilty . . . Confronted by the penal system, the humanist would say: "The guilty are guilty and the innocent are innocent. Nevertheless, the convict is a man like any other and society must respect what is human in him: consequently, flush toilets!" Our action, on the contrary, isn't concerned with the soul or the man behind the convict, but it seeks to obliterate the deep division that lies between innocence and guilt . . . If it were a question of raising consciousness, we could simply publish newspapers and books, or attempt to win over a radio or television producer. We wish to attack an institution at the point where it culminates and reveals itself in a simple and basic ideology, in the notions of good and evil, innocence and guilt.
>
> (Foucault 1977, 227–28)

The group's "attack" was to give prisoners a platform by which they could articulate their living conditions, such as the lack of visitation rights and flush toilets. This act exploded the good/evil and innocent/guilty binaries by exposing and undercutting the assumption that only the good and innocent speak. It is crucial to note that Foucault's goal was not to endorse the outcomes of such speech; he was interested in the fact that such speech, in and of itself, broke the calcification of a unidirectional relation of power between the penal institution and its prisoners. The ability to speak was a reversal in the relations of power that, it just so happened, actually fostered a dramatic improvement in prisoners' treatment at the time.

In a similar fashion, Pompa wants the service-learning experience—through the slamming prison gates and the inside students' reflection on "doing time"—to shake up the binaries that outside students may have about the criminal justice system. This may not, in and of itself, get her students to any predetermined outcomes about the content knowledge under examination, the cultural openness hoped for by liberal advocates, or the political overturning of an "oppressive" system. What it will do is to reveal the "deep divisions" within which and through which we think about the content knowledge, cultural openness, and oppression found within, in this case, the criminal justice system.

Such types of service-learning, though, are difficult to implement and sustain. It is, as such, that the next part of this book explores several alternative models by which to structure, sustain, and institutionalize service-learning within academic programs. For at the heart of a pedagogic strategy of doubt lies a model of critical thinking found within each and every academic discipline: the careful and critical attention to the viewing and understanding of the world through distinct paradigms. The forthcoming chapters thus explore exactly such a "disciplining" of service-learning.

Part II

Institutionalizing Service-Learning in Higher Education

Chapter 4

Disciplining Service-Learning

The service-learning field has been pursuing the wrong revolution. This is the argument in this second part of the book. Namely, service-learning has been envisioned as a transformative pedagogical practice and philosophical orientation that would change the fundamental policies and practices of the academy. But its attempted institutionalization not only faces substantial barriers; the hoped-for political and pedagogical revolution in fact positions service-learning in an uncomfortable double bind. A double bind that co-opts service-learning's agenda such that rather than service-learning changing higher education, higher education will change service-learning.

The following chapters thus argue that a truly transformative agenda may be to create a parallel movement to develop an "academic home"—a disciplinary "home base"—for service-learning. This "disciplining" of service-learning, I will argue, is not the negation of a politics of transformation but the condition of its possibility. Specifically, I put forward the argument that service-learning can be sustained as a legitimate and critical undertaking in higher education only by becoming "disciplined" within the framework of an academic program. By linking rigorous academic coursework with deeply embedded and consequential community-based learning, academic programs embody the connections and engagement desired between institutions of higher education and its local and global communities. What such programs truly offer—to students, institutions, and communities—is a legitimate and long-standing academic space from which to foster a meaningful praxis of theory and practice. It is from within this space that service-learning can truly flourish.

This is because higher education is conservative in that it is an institution that is deliberative and slow to change. If, as I have argued in the first part of this book, service-learning is truly disturbing, and if it is seen as a movement outside of the academy, then it is bound to be diluted and dispelled.

The following chapters thus explore an alternative theoretical framework. Namely, I suggest that we envision service-learning as being "disciplined." By this I mean both the academic notion of becoming an academic program akin to physics and women's studies and becoming disciplined in its critical and self-reflexive examination of its own foundations and enactments. This chapter examines this disciplining through the lens of service-learning as an intellectual movement and in parallel to women's studies, which I consider an exemplary model of an intellectual movement that successfully became incorporated into the academy. Chapter 5 examines how some programs are already institutionalizing service-learning through coherent certificates, minors, and majors. And Chapter 6 takes this thought experiment to its logical conclusion by examining other disciplines and fields—such as community studies—that provide additional insight into the development and integration of community engagement within the academy.

The ultimate goal of this section of the book is to demonstrate that there are legitimate and visible boundaries to the field of service-learning, and that it is thus possible to think of this movement as an academic discipline. It is thus appropriate to begin this chapter by contrasting two seemingly opposed statements.

The first is by Elaine Reuben (1981), at the time the national coordinator of the National Women's Studies Association (NWSA), who argued at the first Summer Institute on Women's Studies at Ann Arbor that the feminist movement "may get lost in our transformation" (quoted in Bowles and Klein 1983, 1). The second is from Wendy Brown, a political science professor at University of California, Berkeley, who argued at the 2002 United Kingdom Women's Studies Network Conference, "Beyond Sex and Gender: The Future of Women's Studies?" that "If we are without revolutionary possibility today, we are also free of revolution as the paradigm of transformation" (Brown 2003, 15).

By the time of Reuben's speech, just barely a decade after the first program began at San Diego State University, there were several hundred women's studies programs scattered across the country, and a fundamental issue brewing throughout that decade was whether the field should be conceptualized as an autonomous academic entity or an integrationist and transformative agenda across higher education. By the time Brown gave her speech, there were close to a thousand women's studies programs and the usual academic accoutrements that accompany such success: dozens of journals and conferences devoted to women's studies issues, multiple stand-alone PhD programs, etc. The fundamental issue on the table at the conference (and over the last decade) was whether women's studies was still possible.

It appears that a simple linear argument about the trajectory of women's studies emerges: a radical social movement intent on changing higher education has instead become co-opted and domesticated to the detriment of both the movement and the peoples meant to be liberated by it. As Martha Nussbaum (1999) phrased it in a review of Judith Butler's work:

> The great tragedy in the new feminist theory in America is the loss of a sense of public commitment . . . [Butlerian feminism] tells scores of talented young women that they need not work on changing the law, or feeding the hungry, or assailing power through theory harnessed to material politics. They can do politics in safety on their campuses, remaining on the symbolic level . . . [But] Hungry women are not fed by this, battered women are not sheltered by it, raped women do not find justice in it, gays and lesbians do not achieve legal protections through it . . . Feminism demands more and women deserve better. (44–45)

There is much, I suggest, that the service-learning movement can learn from the institutionalization of women's studies in higher education. What can be learned is that appearances are deceiving. That Martha Nussbaum is wrong. She is wrong because she conflates the academic with the nonconsequential. She is wrong because she presumes activism is always liberal and always liberatory. She is wrong because, ultimately, she uses disciplinary strategies in an attempt to undiscipline a field (see, e.g., Wiegman 1999, 2005, for critiques of Nussbaum's position). She wants feminism (and theory) her way. But that is the beauty and power of being disciplined. If "we are also free of revolution as the paradigm of transformation," then it becomes possible to listen, to think, to question, to retort, to change, to move on. This, I will suggest, is the truly radical transformation that service-learning should be institutionalizing.

I want to argue that women's studies is a paradigmatic example of how service-learning can and should become deeply and substantively institutionalized within higher education. I want to use the case of women's studies to demonstrate that only by becoming disciplined—by becoming an academic program or departmental unit—can service-learning truly be sustained and nourished in the academy. In fact, I want to suggest that if service-learning does not to some extent become transformed into an academic discipline, it will ultimately become just one more educational reform model scattered haphazardly and ineffectually across the higher education landscape.

I thus propose to begin the dialogue at its most logical conclusion: the service-learning field must embrace and work toward developing an academic program at every Campus Compact member institution. For if the service-learning field does not become disciplined, if it cannot claim the same status as women's studies (or physics, for that matter), it can never discipline itself. And if it cannot discipline itself, if it cannot gain the professional and social legitimacy to control its own knowledge production, develop its own disciplinary boundaries and norms, and critique and further its own practices, it will be undisciplined and unsustainable.

I suggest that service-learning is at a critical juncture in becoming institutionalized, much like women's studies was a quarter century ago. Yet I worry that the service-learning field is not pursuing the strategy I believe most vital to its sustenance: disciplining itself. Thus I put forward this argument. I am very well aware that "we may get lost in the transformation." I therefore begin this chapter by addressing the worries of what it might mean to become institutionalized as an academic discipline, and then show such worries to be unfounded. I then present the case for women's studies as a paradigmatic exemplar for the service-learning movement, with a key insight being that the social movement of feminism was able to transform itself into the academic movement of women's studies. I conclude this chapter by sketching out both the implications and necessary next steps if the service-learning movement is to become disciplined within an academic department—one of the few sustaining and productive means by which to maintain itself within higher education.

Before proceeding, it is critical to make explicit three points. First, the strategic move of developing an academic program cannot be viewed as a radical intervention in and disruption of the present-day trajectory of institutionalizing service-learning. For proceeding apace without change is just as much of a strategic move; not intervening and not changing the trajectory are just as political and have (I suggest) just as perilous outcomes. Second, this chapter is written in fear and trembling exactly because the service-learning field has not yet become disciplined. Wendy Brown (1997) could only write an article entitled "The Impossibility of Women's Studies" exactly because such critique was made possible by the academy's norms of what disciplines are allowed to do. I am worried that not only am I not allowed to write this chapter, but that in doing so, I diminish rather than support the field. Critique should not be so constrained. Third, this argument does not in any way suggest an *either/or* solution. Women's studies scholars long ago realized that they needed to have an autonomous academic home *and* push for the inclusion of feminist perspectives across the academy. Likewise, I believe that service-learning will become successfully institutionalized only when the field adopts a *both/and* strategy. What has been missing, though, is a discussion of how service-learning can become institutionalized as an academic discipline, which is why I first need to address the worries of exactly such a form of institutionalization.

THE CASE AGAINST AN ACADEMIC DISCIPLINE

There are of course numerous legitimate reasons to oppose the idea of transforming service-learning into an academic discipline: e.g., the service-learning field does not have the resources to support two vastly different institutionalization strategies; the idea of an academic program is undertheorized and has minimal models for potential replication. But there are also numerous illegitimate reasons that may be used to oppose this proposal. I want to outflank these to allow the real debate to emerge.

Some may argue that this proposal fosters a "community as lab" phenomenon. "Objective" researchers, it is feared, will descend on "the Other" to test, to probe, and to leave to write up their data in the comfort and safety of their ivory towers. Colonialism and paternalism redux. The first step to dismantling this worry is to note that poorly designed and implemented service-learning already does this. The next step is to make clear that all disciplines create and monitor their own disciplinary assumptions of learning, teaching, and research. Teacher educators ask questions such as "should we lecture in a classroom?"; qualitative researchers debate the ethical dilemmas of fieldwork; economists worry about which statistical models skew the data more than others. Every discipline is a community of scholars worried about particular minor or major crises in their respective fields and subfields.

What is important to note is that worries do not intuitively transfer across disciplines. Economists do not really worry about the efficacy of lecturing. Some economists may of course stumble upon innovative ways to better convey their material, but lecturing versus discussion is not a burning issue in the field. Likewise, teacher educators have little use for the day-to-day minutiae of which statistical models can be used to understand, for example, racial stratification across academic tracks. Put otherwise, one can be a bona fide member of the economics profession without ever grappling between lectures and discussions. One cannot, though, be a bona fide member if one has never grappled with the question of statistical methodology.

Likewise, the means and goals of an academic discipline become one of the fundamental questions in the field. For example, the question of "How much voice should community members have in the partnership?" immediately becomes expanded and problematized: "Whose voices should be heard and whose shouldn't?"; "How should such hearing occur?"; "What does it even mean to hear?" What becomes clear is that there will be (and should be) a spectrum of perspectives about the notions of reciprocity, respect, power, and knowledge production embedded in this extremely complex and multifaceted question. What also becomes clear is that the answer to the question (much less the development of the right questions) necessitates a broad range of academic input. Philosophers, sociologists, anthropologists, and economists, to name just a few, can contribute to this discussion. Moreover, the question about community voice can deeply inform multiple disciplinary discussions across the humanities and social sciences.

The fear of a "community as lab" phenomenon is thus unfounded. To be a member of an academic field devoted to community engagement (as a methodology and focus of inquiry) means that at some point in your academic career you have grappled and hopefully continue to grapple with the question of community voice. If you do not or have not, you may be considered an excellent historian or sociologist or whatever, but you will not be accepted in my club.

A second counterargument to be made is that this proposal will destroy any notion of service-learning as an overarching social-justice vision and framework for action. It will, the argument goes, "ghettoize" service-learning into a theory-laden and activist-poor academic backwater concerned more with publishing and tenure than with real changes in the real world. The first step to dismantling this worry is, again, to note that service-learning is at present not doing much better. For all of the human, fiscal, and institutional resources devoted to service-learning across higher education, there are in fact very minimal on-the-ground changes in the academy, in local communities, or in society more generally. As I have argued, service-learning should not have to bear the burden (nor the brunt) of being the social justice standard-bearer.

The second step in dismantling the fear of service-learning's "planned irrelevance" is to make clear that the rhetoric of service-learning as an overarching social-justice mission falls into the political trap of universalism cited earlier by Fish. To maintain that service-learning can "raise all boats" is to ignore the politically liberal trappings of service-learning as presently conceptualized and enacted. The fear of a loss of vision and mission for service-learning is thus unfounded. It is a presentist and acontextual argument that has created an archetypal narrative of what service-learning is and should be without acknowledging its underlying assumptions and implications. This is, again, not meant to berate service-learning. It is meant to make clear that service-learning is a contested term with no grand universal claims possible about what it is or should be.

A final argument against this proposal is that the service-learning field has come too far to turn back. To plot a new direction, this argument goes, would in essence negate the massive resources, energy, and time devoted to bringing service-learning this far. This is in one respect a valid claim. Unlike women's studies a quarter century ago, service-learning is no longer at the margins of the academy. It is oftentimes at the front and center on many institutions' homepages, marketing brochures, course catalogs, and faculty development priorities. Service-learning, in other words, may have already become too popular to be revisioned.

Yet I reiterate that this is not an either/or scenario. The first step to dismantling the worry about losing "sunk costs" is to point out that costs continue to sink without a firm foundation. I am arguing that the service-learning infrastructure being presently envisioned and enacted cannot sustain itself. At least not in the way that service-learning advocates want it to be. Service-learning cannot become institutionalized without disciplining itself.

And this is the real and deepest fear at the heart of this last worry and within the service-learning field. The second step in dismantling the fear of losing what has been won is to point out that service-learning scholars and activists want service-learning to be all things to all people, yet this is not how higher education works. Becoming disciplined takes time, effort, and perseverance. Once disciplined, there is of course a loss, a trapped quality, but any good scholar will tell you that one must first be trapped by one's own discipline before venturing out to attempt to understand another, much less offer guidance or clarity of vision across the higher education landscape.

The worry about venturing in the wrong direction by embracing such disciplinarity is thus unfounded as well. The development of an academic program would foster all that has already occurred in the service-learning field exactly because it will allow for a disciplined inquiry into the scholarship of engagement and the linkages between higher education and communities.

THE CASE FOR AN ACADEMIC DISCIPLINE

I have shown in the last two chapters the limits of institutionalization as an overarching paradigm. The key insight is that all overarching models of institutionalizing service-learning presume that, by whichever means necessary, service-learning should become an overarching framework for higher education able to be embedded both horizontally across departments and vertically throughout all levels of an institution's pronouncements, policies, and practices. This is nothing less than a grand narrative for higher education-as-service-learning, or more precisely, the notion of thinking about service-learning as a social movement, a politics to transform higher education and society.

I suggest that we must reverse instead the terminology to begin thinking about the politics of transforming higher education and society through service-learning. We must use service-learning not as a political vehicle but as a disciplinary lens—what is referred to as an intellectual movement. I take this distinction from Robyn Wiegman's (2002, 2005) discussion of the future of women's studies. Wiegman argues that Nussbaum's fixation on the sole legitimacy of the former (women's studies as politics) rejects any potential future for women's studies besides the one already discredited:

> For Nussbaum, then, "old feminism" is the authentic and authenticating project of social transformation, one whose real world legibility relies on a tacit privatization of the university as a public political institution in its own right . . . [But] to reclaim the oppositional formulation and to write feminism on the side of women without contradiction or complicity, Nussbaum must forfeit the analytic opportunity that her anger at failure makes intellectually palpable: a consideration of how feminism's institutionalization in the academy has given not only depth and texture but a lengthy archive to a difference previously unperceived between thinking about feminism as a politics and thinking about politics through feminism . . . That future of feminism as a monotheistic politic, narratively equipped with its own fallen angel and moralistically dedicated to pain as an agenda for both knowledge and social change, is finally too allergic to the possibility of any future that second wave feminism has not already imagined. (5–6)

Wiegman's argument is that there never was nor ever will be a feminist movement pure and unadulterated from its own pragmatic and political blind spots and subversions. The feminist movement was and is necessary, but its inability to wipe out gender injustices and sexual discrimination says less about its failings than about the massively interconnected and self-maintaining modes by which such gender bias operates at the cultural and societal levels. This is the insight that Wiegman speaks of as "thinking about politics through feminism," for it allows analytic consideration once it is removed from a moralistic and unitary ideology committed to empowerment pure and simple.

Wiegman is ultimately pointing out that there is no such thing as "authentic" activism. The feminist activism of the 1960s—as second wave feminist scholars pointed out (e.g., Moraga and Anzaldua 1981)—was blind to issues of, for example, class, sexual orientation, third-world status. As Rubin (2005) points out regarding Messer-Davidson's (2002) dismissal of the disciplining of feminism, the "political" is "something other than the mere signifier of a particular social domain or camp affiliation" (246). Political activism and change occurs through many contexts and combinations, and the binary logic of theory/practice dismisses, and thus disallows the power of academic discourse and knowledge production as influential and political in and of itself. Messer-Davidson's claim for a theory/practice divide obscures the much more powerful point that (to take the higher education example) our practices in the street are always already informed by the theories we have come to think with in the classroom.

What I want to suggest is that the service-learning field has not yet gone through its own "second wave" of questioning and critique, and that as it does, it cannot continue to ignore the contradictions and complicities within it; and that with this knowledge, service-learning cannot continue to clutch onto the binary of an "oppositional social movement" embedded within the "conservative" academy. Or more precisely, it can continue to champion itself as a transformative agent of social justice, but only at the price of giving up any analytic opportunity to understand how and why it failed.

So long as women's studies and feminism was (and is) conflated with social activism, it risks being dismissed as yet another form of identitarian politics beholden to the unquestioned uplifting of an essentialized category (e.g., race, ethnicity, gender). What makes women's studies an academic discipline and the gender(ed) subject the mode of inquiry is that its scholarship is able to both look outward and inward. Women's studies can rightly claim to examine an issue (e.g., education, the criminal justice system) through a feminist lens *and* the ability to internally debate and determine what issues are worthy of study, by what modes of inquiry, and to what ends.

It is with this insight that it becomes clear why women's studies was so seriously fractured in the early 1980s by black feminist voices (see, e.g., DuBois 1985; Moraga and Anzaldua 1981) and today by postmodern and poststructuralist critiques. The former attack made vividly clear that women's studies had systematically essentialized and erased notions of race, ethnicity, and class from its purview of study. Likewise, the latter attack makes vividly clear how women's studies has systematically conflated its object of study—the construction of the gendered subject—with modernist notions of progress, liberation, and the unitary self. What is of relevance here is that both attacks have by now become enfolded within the discipline rather than wrenching the discipline apart (though of course the history and outcome of this development might have been very different if women's studies had more tenaciously held onto its singular notion of an activist enterprise).

Women's studies accomplished this by reversing the terminology to make the gender(ed) subject the mode of inquiry rather than use gender as the political project. I suggest that the service-learning field can do likewise by making the community the mode of inquiry rather than by using the community as a political project. This opens up an entirely new model of practice. It becomes possible to use all of the tools of the academy to analyze a very specific and bounded issue. This is the dual meaning of the term "disciplined." There is no doubt that women's studies was disciplined in its institutionalization. It distanced itself from the "street" and from the fervent activism therein; it had to devote attention to bureaucratic maneuverings for funds and faculty rather than for institutional change and transformation; it had to settle for yearly conferences instead of round-the-clock activism. As Messer-Davidow (2002) phrases it, women's studies became routinized.

Yet the appropriation of a Foucauldian terminology of "disciplining" more often than not glosses over Foucault's productive meaning of the term (see Butin 2001, 2002). As Foucault (1997) argued, "We must cease once and for all to describe the effects of power in negative terms: it 'excludes,' it 'represses,' it 'censors,' it 'abstracts,' it 'masks,' it 'conceals.' In fact, power produces; it produces reality; it produces domains of objects and rituals of truth" (194). By becoming "disciplined," women's studies was able to produce the domains of objects and rituals of truth to be studied and recast. As such, I would argue, disciplinary institutionalization is not the negation of politics but the condition of its possibility. It allows scholars, in the safety of disciplinary parameters, to debate and define themselves and their field and how such disciplinary knowledge can inform both the academy as well as local and global communities.

The key insight for this move is that feminism as a social movement became transformed into women's studies as an intellectual movement. This is a lateral move with no easy or short-term pronouncements on the benefits or complications thereof. Likewise, the key is to see the social movement of service-learning as being able to be transformed into an intellectual movement.

Service-Learning as an Intellectual Movement

An intellectual movement may be seen as a specific type and subset of a social movement (Buechler 1993; Mueller 1992; Schneiberg and Lounsbury 2007). While the study of an intellectual movement may have many similar characteristics and points of analysis with social movements—for example, the origins of grievances and tensions; the importance of understanding opportunity structures; networks, mobilization, and collective action dynamics—a key distinguishing feature is that intellectual movements "are collective efforts to pursue research programs or projects for thought in the face of resistance from others in the scientific or intellectual community" (Frickel and Gross 2005, 206).

The traditional model of service-learning that I examined in Chapter 2 through the writings of Andy Furco and the Wingspread statement neatly fits within the notion of a social movement attempting to storm and reform higher education. Yet as I also showed, such a model is inadequate for documenting the complexities of service-learning in the academy and is ultimately not helpful to its sustained impact. But understanding the service-learning movement as an intellectual movement may be.

Frickel and Gross (2005) provide several commonalities among diverse and disparate intellectual movements: (a) that there is a coherent intellectual program that (b) is "contentious relative to normative expectations" within the particular academic domain that is, by its contentious nature, (c) political in the sense of competing for institutional resources (e.g., prestige, tenure-track lines, journals, grant funding), and, (d) "constituted through organized collective action" as opposed to, for example, the ideas, agendas, or projects of any single individual, (e) episodic in that the "birth of a SIM often is marked by the announcement of a bold new intellectual program", and (f) that it varies in its scope and aim (206–8). The service-learning movement as a subset of the larger push for a scholarship of engagement nicely fits within this framework.

The coherency of the service-learning movement may be found within the clarion call of Boyer's *Scholarship Reconsidered* (1990). Boyer's argument for, among other types of scholarships, a "scholarship of application" that genuinely linked academic and community practices synthesized a disparate body of scholarship that had not yet found a singular footing by which to argue for such engagement within the academy. Boyer's articulation provided a clear marker of scholarship that was useful outside of the academy, relevant for more than just scholars, and committed to linkages across and outside of higher education.

The core of such an argument, as I have shown in the first part of this book, is highly contentious relative to the normative expectations within higher education regarding teaching, learning, and research. Critiques of the service-learning movement may be theoretical, empirical, or conceptual (as I outline in the last chapter of this book); irrespective of the critique, of relevance is that faculty and administrators interested in or committed to service-learning find themselves struggling for legitimacy vis-à-vis highly embedded notions of what counts as "academic."

Exactly because the service-learning movement is contentious, it is also political. It is important here to differentiate between a traditional public political movement, something that service-learning is oftentimes associated with, and a movement that is political in the sense of a contestation for resources, position, ability to frame its own narrative, etc. Some aspects of the engaged scholarship movement do indeed have highly political and normative goals—such as the betterment of communities and a social justice vision (e.g., Harkavay 2006). Yet from the perspective of movement theory, these goals are solely coincidental to the extent that there is nothing inherent in an intellectual movement that necessitates a political positioning. Rather, practices such as service-learning and community-based research compete for building space, tenure-track lines, merit-based professional development grants, and administrators' time.

The service-learning field can also be seen as having moved far beyond any single individual's or institution's founding. Stanton et al. (1999) provide an important overview of the top twenty or so advocates for the rise of service-learning through the 1970s. Yet today there is an immense variety of overlapping yet distinct individuals, programs, organizations, and institutions. Such collective action may be seen in the development of attempted "umbrella" organizations: for example, the Higher Education Network for Community Engagement (HENCE) (http://www.henceonline.org/); the International Association for Research on Service-Learning and Higher Education (http://www.researchslce.org/); and the Community-Campus Partnerships for Health (http://www.ccph.info/). The multiplicity of such endeavors may also be seen as the not-yet-realized goal of a truly coherent movement.

The service-learning movement may also be seen as episodic in that it fits neatly within Mullins's (1973) data that an intellectual movement needs one or two decades to stabilize. If one takes the early 1990s as the point of demarcation—for example, the founding of Campus Compact in 1985, the publication of the first "Wingspread Principles of Good Practice in Service-Learning" in 1989, the publication of Boyer's *Scholarship Reconsidered* in 1990, and the federal funding of the Corporation of National and Community Service in 1993—then today's service-learning movement is on the cusp of being a young adult. Interestingly, Frickel and Gross (2005) suggest that a movement's death can be gauged "either by the effective disappearance of the movement from the intellectual scene or by its transformation into a more stable institutionalized form such as a school of thought, subfield, or discipline" (208). Neither of these options has yet happened, though it is exactly the argument of this chapter in specific and this book in general that such a transformation is necessary to avoid exactly the intellectual death presaged by Frickel and Gross's theoretical observation.

Finally, it is possible to argue (though this is least acknowledged in the movement) that service-learning is fundamentally varied in both scope and aim. From the minutia of active learning strategies to a vision of the transformation of higher education, service-learning is played out in multiple forms and with highly multiple goals. I have outlined this variety in the first section of this book and provide additional examples in the forthcoming chapters.

If it is thus possible to mark service-learning as an intellectual movement, then it is also possible to posit a set of propositions concerning its emergence and sustenance. Frickel and Gross (2005) offer several heuristic propositions that align with social movement theory (Morris and Mueller 1992; Swaminathan and Wade 2001): an intellectual movement is more likely to emerge when supported by high-status intellectual actors complaining against what they consider a central intellectual tendency, and when structural conditions provide access to key resources (employment, intellectual prestige, organizational resources); the greater the access to "micromobilization contexts" (e.g., conferences, graduate students, retreats) the greater the likelihood of success; the success of the intellectual movement is contingent upon the movement's ability to frame the movement's ideas in ways that resonate with (potentially) aligned constituents who inhabit other intellectual fields.

The case of women's studies overlays these propositions extremely well. Feminist scholars gained an immense amount of leverage and sustenance from external foundations (such as the Ford Foundation) and internal coalescences that fostered informal and formal workshops, colloquia, conferences, and retreats throughout the 1970s (Stimpson and Cobb 1986; Guy-Sheftall and Heath 1996). Empirical and theoretical work by feminist scholars exploded the myth of scholarly (gendered) neutrality across wide swaths of the humanities and the social and natural sciences (Harding 2004; Lather 2007). And a large number of scholars—irrespective of gender and disciplinary specialization—came to see their own work as linked to and aligned with feminist goals and principles. Feminism as an intellectual movement became transformed into an academic discipline.

The case of the service-learning movement seems less secure. In many respects the service-learning movement is well-positioned in that it is supported by high status figures (academics as well as institutions), and may take advantage of a wide range of institutional, governmental, and academic resources. This strength, though, of access to a wide range of resources outside of the academy, appears to mirror inversely (and thus may in fact contribute to) the minimal progress within higher education per se. Specifically, the access to high-level status and funding has allowed the service-learning field to maintain its status as a "meta" practice above (and thus not grounded in) specific academic practices and protocols. If the service-learning field is going to transform higher education, perhaps it might need to be transformed by it as well.

What It Takes to Become Disciplined

So now what? If it is indeed possible and desirable to become transformed into an intellectual movement, what exactly would that entail? What do we do?

There are some powerful pragmatic opportunities for a move toward the "disciplining" of service-learning. There are currently few journals, conferences, or academic book series devoted solely to the academic study of community-based models of teaching, learning, and research. In one respect this is highly troubling; foundations and book publishers, for example, are reassured by an individual's or institution's track record when evaluating suitability for giving out funds or book contracts. But academic departments committed to a scholarship of engagement do in fact exist (as detailed in the next two chapters). And a multitude of scholars across the humanities and social sciences deal explicitly with community engagement issues. What becomes clear is that there is an opening.

Developing an academic program would also embed it within the very core of the academy. Today's worries about the institutionalization of service-learning would become moot. For example, the *Campus Compact Annual Membership Survey* (2006) cites faculty time pressure, lack of funding, lack of common understanding, lack of funding for work, and faculty resistance as the top obstacles to service-learning on campuses. This is because service-learning is seen as an add-on to all of the other worries, pressures, and constraints on faculty. Yet if there was an academic program, a scholarship of engagement within the community would be the primary task. There would still be time pressures and funding obstacles, but those would simply be part of the job of being a faculty member rather than an additional burden. I would no longer have to worry about whether service-learning was taking time away from my research and potentially preventing my case for tenure. My scholarship of engagement *would be* my research and my case for tenure. Of course I would still have to publish and demonstrate the rigor of my work, but physicists and women's studies scholars have to do that too.

An academic program would also allow the field to move from toolkits to handbooks. As a teacher, I find the "toolkit" metaphor wonderful exactly because it crystallizes the artisan nature of teaching. Huberman (1993) offers a classic picture of the teacher as artisan who is a *bricoleur* (the term is borrowed from Levi-Strauss), managing with a little bit of this and a little bit of that to fashion a truly meaningful and productive classroom environment suited to the particular needs of her students and herself. (This is my own take on the rationale for using the "toolkit" phrasing. I am not aware of how or why Campus Compact chose the terminology that it did.)

But as a scholar I am skeptical. For a "toolkit" metaphor implies that we know the task at hand and we know the tools necessary to achieve it. But as I keep suggesting, service-learning is a multifaceted and contested term able to be used in distinctly diverse ways. If all you have is a hammer, then all problems look like nails; likewise, if all you have is a toolkit, all problems look fixable.

A handbook, alternatively, places questions rather than answers as central to the movement. A handbook allows a field to question itself, to systematize itself, and to remake itself. The most recent *Handbook of Research on Teaching* (2001), for example, was years delayed (in part) because of questions of how (or if) one could incorporate the postmodern move within educational research. More broadly, as David Downing (2004) says of every anthology of theory, all must confront "an institutional double bind: On the one hand, many of the theoretical essays included in the anthology tend to challenge, cross, or disrupt disciplinary borders; on the other, anthologizing itself cannot avoid its essentially disciplinary function" (342). While there is no easy way out of this, the answer surely does not lie in some type of action "outside" of disciplinary specialization. Rather, and again with Foucault (1997), I suggest that the service-learning field must find new "trump cards" by which to reframe academic discourse.

I want, in conclusion, to return to my opening quotes of this chapter and to my suggestion of developing academic programs at every Campus Compact institution. I am not suggesting that such programs are the silver bullet to institutionalizing service-learning across higher education. It trades in one set of worries for another. What I am suggesting, though, is that this new set of worries would be much less worrisome than the present one. Getting "lost in our transformation" by developing complementary academic homes may be one of the only possibilities for sustaining powerful service-learning models in a higher education context that is "without revolutionary possibility."

Finally, I should note that this is no ode to some safe, motherly, and archetypal "home." This is an acknowledgment that knowledge is disciplined by the particularities and specificities of mundane and totalizing structures, policies, and practices. Disciplines and disciplinary knowledges are forged and crafted by (to name but the most obvious) conference papers, journal articles, book series, philanthropic funding, research institutes, job openings, tenure-track faculty lines, *Chronicle of Higher Education* articles, and external reviewers. And as they say about the lottery, "you gotta be in it to win it."

As an academic program or department, we would have to worry about tenure-track faculty lines and resource allocations vis-à-vis other institutional funding priorities. We would have to worry about developing graduate programs to train a new cadre of academics not beholden to other departments' norms and preconceptions. We would have to worry about the rigor and quality of our courses. We would have to worry about our value to the communities we work with and for.

We would have to worry about how to articulate a cohesive and coherent vision of what we are and should be within higher education and to society at large. We would have to worry about whether it was even possible or worthwhile to articulate such a vision.

These worries, it may be argued, are pedestrian and insignificant compared to what is now being discussed. But I beg to differ. Yes, service-learning may be lost in the transformation. But if we are truly free of revolution as the paradigm of transformation, an entire new field of possibilities opens itself up. Service-learning may no longer claim that it will change the face of higher education. But women's studies does not do that either anymore.

Instead, women's studies scholars carefully and systematically elaborate how feminist perspectives are slowly infiltrating and modifying the ways specific disciplines and subdisciplines work, think, and act (see, e.g., Stanton and Stewart 1995). This is not radical and transformational change. This is disciplined change. It is the slow accretion, one arduous and deliberate step at a time, of contesting one worldview with another. Some of it is blatantly political. Some of it is deeply technical. Much of it is debatable, questionable, and modifiable. Just like any good academic enterprise.

And it is this that is truly transformational. What I am proposing will take immense time, funding, and talent. The ultimate directions and outcomes are far from clear. But the immediate path is obvious: We should think and act like good scholars.

Namely, we should debate and discuss this proposal in multiple forums and venues and with multiple stakeholders; we should garner funding from our institutions, from federal grants, and from private foundations to develop pilot projects; we should set up an internal working group within Campus Compact to explore the feasibility and action steps necessary to develop this agenda; we should form more journals; we should organize additional conferences; we should question why we are doing this and, once we are doing it, assess what we have accomplished and failed to accomplish; we should look to our colleagues in other disciplines to help us understand what we are doing, what we should be doing, and why what we are doing differs from what they are doing; we should begin to map out what such a field encompasses, what it does not, and why; we should begin to articulate how such an academic program should function, how it should not, and why.

Much of this is already being done in different parts of the service-learning movement. What I am thus suggesting, to put it simply, is that we should become disciplined.

Chapter 5

Majoring in Service-Learning?

So what would a "disciplined" service-learning look like? How does the philosophy and pedagogy of service-learning become transformed from an all-encompassing movement to an academic discipline? What exactly would a student study? What are the guiding texts, the underlying assumptions, the modes of practice that would guide such an academic field? Or to put it in "undergraduate-speak": "Can I major in service-learning?"

Service-learning has long been theorized and enacted as both a pedagogy and a philosophy. It has been seen as an instructional strategy to support active learning, as a lens of engagement and collaboration with local and global communities, and as a social-justice worldview that has fostered respect and reciprocity amongst diverse constituents across town/gown barriers (Bringle and Hatcher 1995; Kendall 1990; Rhoads and Howard 1998). This is service-learning-as-method, that is, service-learning as a means to a particular goal, be it better instruction, a deeper sense of cultural competence, or social equity. But as I showed in the first part of this book, such conceptualizations of service-learning-as-method are self-limiting and self-undermining. And as the previous chapter suggested, there is a theoretical opening by which to embrace a disciplinary perspective of service-learning.

Moreover, above and beyond the theoretical problematics and openings lies the concrete reality that there has been a slow accretion of programs in higher education that award certificates, minors, and/or majors in service-learning. Vaughn and Seifer (2004/2008), for example, have documented twenty-five such programs, and my own analysis (to be detailed in the following chapter) finds numerous others. This on-the-ground reality reveals an important and unacknowledged dilemma within the service-learning movement. Namely, if service-learning is positioned as an academic discipline, what exactly is one being "disciplined" into?

Scholars of higher education have suggested that academic disciplines teach undergraduates specific and distinctive habits of thinking (e.g., Baxter-Magolda 1999; Becher and Trowler 2001; Lueddeke 2003). Ways of looking at and studying the world are radically different in, for example, economics, physics, anthropology, and women's studies. So what about service-learning?

Such a question highlights a fundamental tension within the service-learning movement. Namely, majoring in service-learning is a fundamentally different situation. A student can major in economics, but not in quantitative research; in education, but not in cooperative learning; in women's studies, but not in feminism. A major (or a minor) presumes that there is an explicit, coherent, and bounded field of knowledge. All disciplines have, of course, at one point or another struggled to differentiate and legitimate themselves vis-à-vis established disciplinary fields. Anthropology and psychology did so at the turn of the twentieth century, and more recent (and successful) examples include women's studies, area studies, and black studies in the 1960s and '70s (see, e.g., Stanton and Stewart 1995; Rojas 2007).

This chapter thus engages the question of "Can I major (or minor) in service-learning?" and its multiple political, pragmatic, and philosophical implications and permutations. It does so as a means to highlight the convergences and distinctions across service-learning programs in higher education and what this might mean for the state of the service-learning movement in higher education. It is an attempt to show that there is indeed a coherent (though far from stable) "field" of service-learning, which, in turn, suggests that service-learning can indeed fruitfully be disciplined.

THE STATE OF THE ACADEMIC SERVICE-LEARNING FIELD

Vaughn and Seifer (2004/2008) document twenty-five programs that focus on "service-learning, leadership, and/or community service" through majors, minors, and certificate programs (1). These programs, they argue,

> tend to focus in one of three areas: 1) A minor or certificate earned by engaging in community service and service-learning activities,

> 2) A minor or certificate earned by learning about the theoretical roots of service-learning and engaging in service-learning activities and 3) A minor or certificate focused on leadership and social change, for which a requirement is engagement in service-learning activities. A small number of higher education institutions offer a major area of concentration that focuses on service-learning. (1)

Vaughn and Seifer suggest that the commonalities in all of these programs are their commitment to students' engagement in service-learning activities. But the deeper issue that is left unexamined is the larger point of how and why some programs have developed their programs as they have. Constructing a certificate, minor, or major program in higher education is a slow and deliberate process that requires multiple stages of academic review, from curriculum committees to Provost's approval to other departments' consultations. To construct a program focusing on service-learning, as such, requires a scholarly articulation of the focus and content of such a program and how it differs from or expands upon existing academic programs.

Put otherwise, I suggest that there is a fundamental distinction between programs that simply use service-learning and those that examine service-learning as a locus of study. The former is the traditional way of thinking about service-learning as a quasitransparent method that can be added to most academic programs of study as a means to enhance achievement of predefined outcomes (be it content, cultural competence, or social justice). Alternatively, programs that position service-learning as a locus of study do not just use service-learning; they deliberately foster examination of the practice. This entails academic study of, for example, the assumptions and positionality of those doing the service, the means and outcomes of the service being done, and the hermeneutics of understanding how the service done links to and informs the specific academic content being studied.

I suggest that Vaughn and Seifer have constructed a list of programs potentially in the latter category (that use service-learning as a locus of study) within the theoretical confines of the former category of service-learning-as-method. Yet to conflate the two distinctive modes is to prevent the reconceptualization of how we might begin to rethink service-learning. It is thus important to reexamine existing programs within the frame of service-learning as an academic discipline.

Vaughn and Seifer (2004/2008) list twenty-five programs that offer certificates, minors, and/or majors that recognize service-learning. To these I have added four others, based on my own research on community studies programs (to be detailed in the next chapter). Additionally, two programs (Emory and Henry College and Providence College) had both a major and minor; I counted them as separate programs as such in that they had distinct and different sets of requirements. Table 5.1 provides an overview of these thirty-one programs.

Two questions drove the analysis of these programs: "How were the programs structured?" and "What content did the programs teach?" An analysis of institutional, departmental, and program Web pages and relevant documents (e.g., requirements within the institution's academic

Table 5.1 Overview of all programs with concentrations, minors, and/or majors in service-learning

Institution	Program Type	Title of Program
Assumption College, MA	Minor	Community Service Learning
Bryant University, RI	Major	Sociology and Service-Learning
CSU-Monterey Bay, CA	Minor	Service Learning Leadership
College of St. Catherine, MN	Minor	Civic Engagement
Colorado School of Mines, CO	Minor	Humanitarian Engineering
DePaul University, IL	Minor	Community Service Studies
Emory & Henry, VA	Major	Public Policy and Community Service
Emory & Henry, VA	Minor	Public Policy and Community Service
George Mason University, VA	Major	Concentration in Public and Community Engagement
Humboldt State University, CA	Minor	Leadership Studies
Indiana University, IN	Minor	Leadership, Ethics, and Social Action
Kansas City Art Institute, MO	Certificate	Community Arts and Service-Learning
Murray State University, KY	Certificate	Service Learning Scholars certificate
Northwestern University, IL	Certificate	Certificate in Service-Learning
Portland State University, OR	Minor	Civic Leadership
Providence College, RI	Major	Public and Community Service Studies
Providence College, RI	Minor	Public and Community Service Studies

(*Continued*)

Table 5.1 Continued

Saint Louis University, MO	Certificate	Service Leadership Certificate
San Jose State University, CA	Minor	Service-Learning
Salt Lake Community College, UT	Certificate	Service Learning Scholars Program
Slippery Rock University, PA	Minor	Community Service and Service-Learning
SUNY-Stony Brook, NY	Minor	Community Service Learning
University of Baltimore, MD	Major	Community Studies and Civic Engagement
UCLA, CA	Minor	Civic Engagement
University of Kansas, KS	Certificate	Service Learning
University of Massachusetts-Boston, MA	Major	Community Studies
University of Missouri, MO	Minor	Leadership and Public Service
University of North Carolina-Chapel Hill, NC	Certificate	Public Service Scholars Program
University of San Francisco, CA	Minor	Public Service
University of Wisconsin-River Falls, WI	Certificate	Service-Learning
Vanderbilt University, TN	Major	Concentration in Community Leadership and Development

catalog) allowed an answer to the first question; a content analysis of these programs' syllabi (gathered through e-mail solicitations to each program) allowed an answer to the second question.

Given the inductive nature of the data—that is, arising from each specific academic program—it is not possible to make any formal claims of comparability across programs. Rather, what holds these programs together as worthy of analysis is instead their self-definition as coherent academic programs. Such self-definitions of course create multiple limitations to the findings: it is unclear to what extent programs with the same or similar names have similar conceptualizations of their practices, and numerous programs that do not self-define in this way or use slightly different terminology to define themselves may not have been included. Yet such a limitation can also be seen as a manifestation of the very way that the service-learning field has organized itself and the consequences thereof. A final limitation to this analysis is that the findings may be limited both by the small number of programs analyzed and by the fact that the unit of analysis was the program documents (for example, departmental websites and syllabi) and not actual faculty practices or student beliefs.

Figure 5.1 provides an overview of the descriptive findings of the programmatic structures of these thirty-one programs. The thirty-one programs include eight certificate programs (26 percent of the total), sixteen minors (52 percent of the total), and seven majors (22 percent of the total).

Of particular interest is that almost all programs—twenty-two out of twenty-six (85 percent)—had some type of formal academic requirement that focused on the theory and/or practice of the service-learning experience. For many programs this was an introductory course. Other programs required a capstone experience that took the form of a senior seminar or an individualized project such as a portfolio, senior thesis, or a community-based project or field experience. Twelve programs (46 percent of all programs) required both an introductory course and a capstone experience. This split among programs—that is, between programs with formal "book-end" structures of introductory and concluding requirements and programs without such formalized requirements—was in fact quite pronounced and obvious. It may thus be helpful to describe two programs on either side of this divide to make visible the differences.

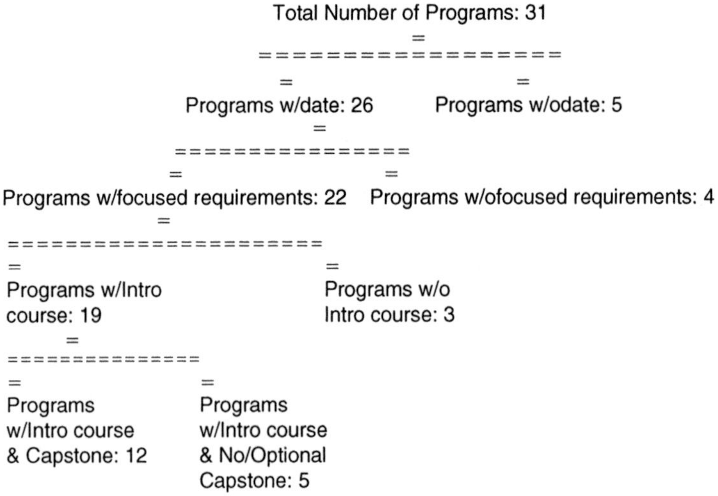

Total Number of Programs: 31

Programs w/date: 26 Programs w/odate: 5

Programs w/focused requirements: 22 Programs w/ofocused requirements: 4

Programs w/Intro course: 19 Programs w/o Intro course: 3

Programs w/Intro course & Capstone: 12 Programs w/Intro course & No/Optional Capstone: 5

Figure 5.1 Overview of program structures

The University of Kansas, Kansas, offers a Certification in Service Learning program through its Center for Service Learning (University of Kansas 2008). The certification is gained once a student completes four distinct components: a classroom experience that requires the completion of an approved course that incorporates service-learning; directed readings that include three articles and a report to be used "as a resource in the final reflection paper" (1); an independent project that may be another course with service-learning, an alternative break, or additional volunteering or leadership experience; and reflection that may be fulfilled either through attending "two one-hour reflection sessions" (2) or a written paper of eight to ten pages that focuses on the "what?" "so what?" and "now what?" of the students' service-learning experiences.

The University of North Carolina (UNC)-Chapel Hill program is structured in a similar fashion. Students in the Public Service Scholars program are required to complete 300 hours of service, take an approved course at the college with a service-learning component, complete four "skills training" sessions, and write a "senior portfolio" that is a reflective 750-word essay (UNC 2008). The "skills training" component is meant to foster skill sets such as advocacy, ethics, and organizational leadership, and may be fulfilled by attending a wide variety of conferences, workshops, or courses. Similarly, the approved course can be from across the college or may take the form of an alternative spring break trip or an independent study.

What is indicative of these programs is the lack of a coherent and deliberate engagement with what constitutes service-learning, which is viewed as a given, a taken-for-granted process and product that needs no guided deliberation or debate. Both programs allow students to take any institutionally approved course with the moniker of service-learning. Both programs allow flexibility in accomplishing such tasks, with immense scope ranging from an alternative spring break to leading a student group. And both programs allow students to write, without the context of a course or professor, a short reflection on their experiences.

The issue here is not one of quality or rigor (though, in fact, both of these programs seem to call such issues into question). Rather, the issue here is the lack of a deliberate and sustained engagement with what it means to be engaged in the process of service-learning. Let me provide two other examples—Northwestern University's certificate program in Illinois and the minor in leadership, ethics, and social action at Indiana University, Indiana—that mirror more structured programs.

Northwestern University's Certificate in Service Learning program requires students to take five courses, do 100 hours of community service, attend a set of reflective seminars that occur on a biweekly basis throughout the five-academic-quarter sequence, and a capstone project (Northwestern 2008). The required courses include two academic ones—Introduction to Community Development and Leadership and Community Decision-Making—as well as an approved course with a service-learning component and two independent studies that function as the capstone experience. The capstone project, either in groups or for individuals, should "have relevancy to the sponsoring organization's mission and goals" (3), and is presented to the larger academic community in a public forum.

Indiana University's minor in leadership, ethics, and social action (Indiana 2008) provides another common model for programs with minors and majors. All students must complete a required introductory course, choose three electives within concentrations of ethics, social organizations, and social action, and complete a capstone project and seminar. All of the courses incorporate service-learning, and the capstone project includes an eight to ten page paper, a reflective journal, and a public presentation that is linked to the student's service component and the faculty member's readings.

A comparison of these two programs with the former two reveals several key distinctions. Specifically, these latter two programs have—above and beyond the required field-based service-learning and discipline-based coursework that incorporates service-learning—deliberate coursework and sustained inquiry that focuses on the topics and issues inherent within the service-learning experience. The required courses in both the programs provide all students entering the programs with a coherent set of common texts, perspectives, and analytic tools by which to make sense of their future coursework and field-based experiences. Moreover, both programs culminate in a capstone experience that has a public component and integrates the student's previous experiences.

The distinction between the programs—University of Kansas and UNC-Chapel Hill versus Northwestern University and Indiana University—lies in a fundamentally different perspective on how to think about and thus engage with the practice of service-learning. The University of Kansas and UNC-Chapel Hill, much like the other programs without formal coursework and capstone requirements, function with service-learning as a practice simply to be carried out. The other programs view service-learning as something to be examined.

This formal examination may be viewed as the operationalization of my conclusions in Chapter 3 that there is no such thing as service-learning without remainder. If service-learning is understood to be a socially consequential, culturally saturated, politically volatile, and existentially defining practice, then it is always—from before it is even begun—impacting and disrupting the academic content meant to be taught. Thus, as Boyle-Baise et al. (2007) have argued, we must be prepared to reverse our terminology and learn about service before we can fruitfully do service-learning.

The formal requirements (or lack thereof) are thus the embodiment of a program and institution's implicit and explicit assumptions and visions of what constitutes service-learning. The formal requirements (or lack thereof) may also, of course, be dependent on the particular and idiosyncratic cultural and historical contexts and policies of each particular institution and department. It is thus relevant to determine what (if anything) impacts a program's structure. Is it the size or type of the institution or the type of student body? Is it the institution's stated level of commitment to service-learning?

To analyze this distinction more formally, it was thus necessary to determine whether alternative explanations—above and beyond the program structure—impacted how service-learning programs were organized. An exploratory correlational analysis was thus conducted. Each of the programs was coded according to a host of variables (with data found through the Carnegie Foundation's classifications and through institutional and departmental websites): private/public, size of institution, traditional Carnegie classification, residential status of students, and whether the institution had gained the Carnegie "community engagement" voluntary classification. A Pearson correlation was run to determine whether a program's formal requirements correlated to any of these variables. The requirement of a capstone experience had no statistically significant correlations. The requirement of an introductory course had a single statistically significant relationship: to program type. Table 5.2 provides a summary of the results.

This finding is surprising. It suggests that an institution's size, the achievement of the Carnegie Foundation's "community engagement" classification, public or private status, or the residency status of its student population does not matter to the requirement of introductory coursework. One could imagine a host of rationales for why institutions would require formal introductory coursework: institutions already committed to strengthening service-learning (as documented by proxy variables of the Carnegie classification) or institutions that were smaller and/or more focused on serving traditional

Table 5.2 Pearson correlations of institutional variables and service-learning program types

		Required SL Intro Course
Required SL Intro Course	Pearson Correlation	1.000
	Sig. (2-tailed)	
	N	28
Type of Program	Pearson Correlation	.671**
	Sig. (2-tailed)	.000
	N	28
Community Engagement classification	Pearson Correlation	.050
	Sig. (2-tailed)	801
	N	28
Minor is linked to major	Pearson Correlation	.175
	Sig. (2-tailed)	.372
	N	28
Public or Private	Pearson Correlation	.254
	Sig. (2-tailed)	.201
	N	27
Size of College	Pearson Correlation	–.194
	Sig. (2-tailed)	.322
	N	28
Residential Type	Pearson Correlation	.075
	Sig. (2-tailed)	.717
	N	26
Carnegie classification	Pearson Correlation	–.358
	Sig. (2-tailed)	.061
	N	28
Required SL Capstone	Pearson Correlation	.272
	Sig. (2-tailed)	.178
	N	26

** Correlation is significant at the 0.01 level (2-tailed).
* Correlation is significant at the 0.05 level (2-tailed).

student populations (as documented by proxy variables such as size, residential type, public/private, and Carnegie classification).

Instead, the dividing line falls squarely and cleanly across program type. Table 5.3 clarifies this by showing that required introductory coursework is only required in one certificate program (Northwestern University's, as described above), whereas nineteen out of twenty-one programs with minors or majors require such an introductory course.

In one respect this is natural. Minors and majors are the programmatic structures by which disciplinary fields signal that there is a coherent and distinctive body of knowledge. And an introductory course is the standard means by which a field thus begins to introduce

Table 5.3 Cross-tabulation of type of program by program requirement of introductory course

		Required SL Intro Course		
		No	Yes	Total
Type of Program	Certificate	6	1	7
	Minor	2	12	14
	Major	0	7	7
	Total	8	20	28

students to the nomenclature, issues, and goals of its distinct body of knowledge and ways of viewing the world. Yet this realization also draws two immediate responses. The first—which will be examined at greater length in the concluding section of this chapter—is the meaning and implications of (a lack of) introductory coursework, particularly at the certificate level. The second—which will now be taken up—is the question of what exactly it is that I am majoring (or minoring) *in* when I enter a program in service-learning?

MAPPING THE CONTENT OF THE SERVICE-LEARNING FIELD

It becomes apparent that program structure is linked to at least an introductory course that is potentially linked with a formal capstone requirement. Such a formalized "book-end" structure constitutes a standard and traditional model. The question of what actually is taught within these programs now becomes relevant. Namely, what is it exactly that I study when I pursue a formalized path within the service-learning field?

All nineteen programs with introductory coursework were contacted by e-mail, and nine responded by providing the current syllabi of their introductory courses. The high response rate (47 percent) and the lack of bias in the type of program that responded suggests a strong degree of validity to the data and implications below. One program (Portland State University, OR) sent two different syllabi (from two different professors) for the same introductory course. The following analysis is thus based on ten syllabi of courses as shown in Table 5.4.

An initial finding is the wide variance in the focus and disciplinary leanings of the courses. Three of the courses (DePaul, Indiana, Missouri) focus on community involvement; two of the courses (Northwestern, Vanderbilt) focus on community development; two

Table 5.4 Syllabi of required introductory coursework

Institution	Title of Course
DePaul University, IL	CSS 201 Perspectives on Community Service
Emory and Henry, VA	PPCS 100 Intro to Public Policy and Community Service
Indiana University, IN	LESA 105 Beyond the Sample Gates
Northwestern University, IL	SESP 202 Introduction to Community Development
Portland State University, OR	PA 411 Foundation of Citizenship
Providence College, RI	PSP 101 Introduction to Service in Democratic Communities
University of Missouri, MO	HCCIP Honors College Community Involvement Program
University of San Francisco, CA	POL 118 Intro to Public Administration
Vanderbilt University, IN	HOD 2600 Community Development Theory

of the courses (Portland, Providence) focus on issues of citizenship/democracy; and two of the courses (Emory and Henry, University of San Francisco) focus on issues of public policy. While an overarching theme throughout might be "community" and "democracy," neither of these terms are analytically distinct enough to presume a focus of study for the service-learning "field."

A sampling of each course's objectives and statements of purpose make clear, in fact, that each course is distinctly positioned within a specific literature and academic frame of analysis. The course overview of Northwestern University's introductory syllabus, for example, states that "this course will examine both historic and contemporary community building efforts, paying special attention to approaches that were shaped by Chicago [i.e., Jane Addams, Saul Alinsky]." DePaul University's introductory course, alternatively, states: "In this course, we will explore together the uniquely American perspective on community service beginning with its historical foundations in the U.S. to recent attention on national community service programs." And Vanderbilt's syllabus states that this course "is designed to provide an introduction to the field of community development by exploring diverse forces that influence urbanization and community development processes."

Many of the syllabi do in fact explicitly address issues of citizenship, service, and community through the lens of leadership and/or public policy. Indiana University's course states that "in this class you will learn about acting in public life by participating in your community";

Portland State University's course states that "in this course we will examine the place and function of leadership in democratic societies and the ways in which people put conceptions of civic responsibility into practice"; and Providence College's course states that "the course will focus on three concepts that are central to our 'studies': service, community, and democracy (which incorporates issues of equality and social justice)."

The frame and focus of service-learning is thus deeply interdisciplinary, as topics such as leadership, citizenship, and public policy can be approached from distinct and distinctive analytic perspectives. To approach the notion of community from the grounding of community development is fundamentally different from approaching it from a perspective of public policy and leadership or history.

What does stand out is the emphasis on academic scrutiny and critique. While this may appear obvious in the setting of an academic course, it is far from obvious if one returns to the examples of University of Kansas and UNC-Chapel Hill. What all ten syllabi articulate clearly is the need to carefully and critically engage with the specific issues under examination. Thus Indiana University's syllabus continues that the "acting in public life by participating in your community" is fundamentally "a foundation in the organizing skills they [the students] will need for the capstone project in the minor . . . Most importantly, you will be encouraged to follow your own questions to a deeper level"; likewise, Portland State University's syllabus continues by stating that "students will be challenged to examine the promise and challenges of community building and leadership development in the context of our evolving democratic society," and Providence College's syllabus informs students that "each concept [service, community, and democracy] will be explored from three different perspectives . . . (1) concrete and practical experiences or 'case studies' . . . (2) critiques of the concepts (i.e., the challenges/criticisms posed for those interested in promoting service, community, or democracy); (3) good ideas and 'best practices.'"

An academic course is thus the site for the careful examination of specific concepts and ideas, with such examination entailing scrutiny and critique. Such scrutiny is most likely balanced by additional readings and discussions of "best practices" and positive implications of the specific issue under examination; nevertheless, the key point here is that students are exposed through readings, lectures, and the instructor's setup of the course to probe the potential limits, contrary perspectives, and unintended consequences of the so-called reform in question whether it be leadership, citizenship, or service.

This can be clearly seen when one does a deeper content analysis of the syllabi. Namely, both the readings and course assessments in each class presume a critical and analytic stance on the particular course topics under investigation. While there are very few course readings in common (an issue addressed in the concluding section of this chapter), those in common do suggest a core of critical perspectives on notions of citizenship and service. Likewise, the means of assessment across courses is typical and skewed toward gauging the formal analytic competence of its students.

The required readings in each syllabus were coded by both author and text. Edited volumes as well as coauthored texts were divided into sections by individual authors and excerpts to capture the diversity of required readings. A total of 162 unique authors representing 173 unique texts were found across the ten syllabi. (Different texts by the same author were used across courses, accounting for more texts than authors.) The authors and readings ranged from the famous and well-known (e.g., Martin Luther King, Jonathan Kozol, the Bill of Rights) to the highly specialized (e.g., Tracey Smith's "Trashing Appalachia"; Murphy and Carnevale's "The Challenge of Developing Cross-Agency Measures"). A startling finding was that just over 10 percent of the authors could be found in more than one syllabus, only four authors could be found across three syllabi, and only a single author—Robert Putnam—was found in four syllabi. Table 5.5 provides a synopsis of every author who was found in more than one syllabus.

The most commonly used authors and readings—Putnam's *Bowling Alone*, Benjamin Barber, and De Tocqueville's *Democracy in America*—appear natural choices for issues of citizenship, service, and democracy. Yet of particular interest is that most of the other common readings have a critical edge toward examining these very same notions. The readings by Jonathan Kozol, Barbara Ehrenreich, Thich Nhat Hanh, David Hilfiker, and Paul Loeb all question, to one extent or another, our societal will and ability to serve all citizens equally and equitably. Likewise, the readings of Peggy McIntosh, bell hooks, Michael Ignatieff, Ivan Illich, and John McKnight question whether the very act and desire for equal and equitable citizenship is possible given the deeply gendered, racialized, and classist society we live in.

In fact, several of the readings—such as Illich, hooks, and McKnight—seemingly undercut the ability of community engagement and service-learning to be a powerful change for the better. And such readings are far from unique. Other syllabi used texts by, among

Table 5.5 Frequency count of authors and texts within service-learning syllabi

	Number of Occurrences (total N = 10)	Title of text, if same
Putnam, Robert	4	Excerpts from *Bowling Alone*
Barber, Benjamin	3	Diverse readings
De Tocqueville, Alexis	3	Excerpts from *Democracy in America*
Kozol, Jonathan	3	Diverse readings
McIntosh, Peggy	3	"White Privilege"
Ehrenreich, Barbara	2	Diverse readings
Hanh, Thich Nhat	2	Diverse readings
Hilfiker, David	2	Excerpts from *Not All Of Us Are Saints*
hooks, bell	2	Diverse readings
Ignatieff, Michael	2	Diverse readings
Illich, Ivan	2	"To Hell with Good Intentions"
Kretzman, John	2	Diverse readings
Loeb, Paul	2	Diverse readings
LeGuin, Ursula	2	Diverse readings
McKnight, John	2	"Why Servanthood is Bad"
Morton, Keith	2	Diverse readings
Neusner, Jacob	2	Diverse readings

others, Max Weber, Hannah Arendt, Stanley Fish, Harry Boyte, and Paulo Freire to make similar points that the practices and goals of a particular class—be it leadership, service, or community development—are never straightforward, self-evident, or naturally good. As with any academic endeavor, the limits and boundaries of the object under examination had to be probed and tested.

Such an academic stance can also be seen in the academic requirements and assessment practices of the syllabi. Unlike with the required readings, there was a high degree of commonality for how students would be assessed for completing course expectations and outcomes. Table 5.6 provides a summation of a simple frequency count of assessment requirements in a syllabus that directly impacted a student's grade.

None of the assessment mechanisms are in and of themselves noteworthy. Grades based on in-class participation, examinations, and reflective and analytic papers are standard fare in the undergraduate classroom (Angelo and Cross 1993). Which is, in fact, what is noteworthy. Namely, these courses presume that a course that fulfills the program requirement for service-learning be structured much like any other academic course. The emphasis on traditional

Table 5.6 Frequency count of modes of assessment of coursework in service-learning syllabi

Assessment mechanism	Number of Cases (total N = 10)
In-class participation	9
Exam(s)	7
Reflection paper / journaling	7
Analytic paper	6
Group project	5
Community service requirement	2

academic requirements makes clear that students must hold to the expected rigor and expectations of the particular instructor. Such service-learning courses are thus not simply and solely about talking and reflecting and dreaming of a better world. They are engaged in the all-too-common practice of higher education of critical thinking, careful attention to detail and data, and respect for expert knowledge.

IMPLICATIONS FOR THE FIELD

To return to the question at the start of this chapter, it appears that, yes, it is possible to major in service-learning. In fact, gaining a minor or major in service-learning appears to be very similar to gaining a minor or major in any other interdisciplinary field such as women's studies, black studies, or Jewish studies. I explore this idea in detail in the next chapter. For now, though, I simply want to suggest that this point of commonality may be the service-learning field's greatest strength and weakness.

The academic program, and specifically the academic department, is the cornerstone of the workings of higher education. Service-learning programs so structured have the telltale signs of academic legitimacy: standard academic coursework and readings within the standardized format of introductory coursework and capstone requirements. Each program certainly approaches the content matter through its own disciplinary particularity (e.g., political science, history, etc.). But this is a given (much like with any other interdisciplinary field) in that service-learning is not sui generis. All academic analyses and examinations build upon existing literature strands and theoretical contributions.

To major or minor in a disciplinary field is to undertake a programmatic study—constructed by a particular group of faculty within the sphere of a more or less constrained body of knowledge that has come to be constituted as a field. The actual contours of a field may of course be constantly up for debate, and even guiding principles may come under question (one need only think of psychology's split from philosophy at the turn of the twentieth century). Yet a major or minor so constituted presumes a consciously structured and sequenced program.

Recent examinations of the constitution of an academic program (e.g., AAC&U 2007) have reinforced the necessity for such programmatic structures to the quality of undergraduate study. As the AAC&U (1992) noted in a set of reports across the social sciences: "Strong programs help students develop the capacity to use the methods and perspectives of the discipline(s) in framing questions and in developing increasingly sophisticated analyses of those questions. Recognizing that these capacities develop over time, these programs create curricular structures that provide students with opportunities to revisit issues that they have met in prior courses" (12). This structure, moreover, should be cogent: "The program should be organized around a careful plan that views it as a coherent whole rather than as simply a collection of courses" (6).

The vast majority of service-learning programs that are planned as minors or majors do in fact have such a formalized and coherent structure, beginning with an introductory course and building up over the semesters with field-based experiences, reflective opportunities, and culminating in a capstone experience. Such a tripartite structure—introductory coursework, field-based experience, capstone—surely serves as a framework for the field.

But the bigger picture that the research of this chapter demonstrates is that the majority of existing service-learning programs are in fact not structured in this way. Service-learning programs that are not structured in minors or majors (constituting the majority of those listed in Vaughn and Seifer's [2004/2008] compilation) have minimal means by which to shape the academic and social narratives of what constitutes service-learning and community engagement. The extreme flexibility of coursework options, self-guided reflection, and stand-alone field-based experiences may facilitate a shallow institutionalization. But it does so at the expense of being able to engage with the issues raised, the goals attempted, and the means used. This is the operationalization of Wiegman's (2001) argument of the last chapter: that clinging to the notion of a social movement (be it feminism or service-learning) as a change agent without remainder ultimately undermines the analytic opportunity to engage with its own limits and possibilities. Service-learning programs without structured academic requirements thus function as little more than placeholders for already formed and already limited notions and beliefs about the value of community engagement.

While the next chapter explores how some strongly structured programs function and look, it is possible to develop and build upon the findings of existing programs. First, existing programs offer students a sequential and integrated curriculum that, minimally, had an introductory foundational course and concentrations or tracks that aligned with existing programs (e.g., sociology, anthropology, political science, and public policy). Students were introduced to issues surrounding engagement with the community (be it in the form of service, development, or organizing) and built upon such knowledge through future coursework.

Second, many minors and majors made use of particular methodologies that structured how students came to examine such community engagement. Northwestern University's certificate program required a course called "Leadership and Community Decision Making" as a means to follow up on the introductory course's focus on community development. Vanderbilt University required a course called "Action Research and Program Evaluation" that built upon the introductory course's focus on community development theory. It is acknowledged that, in theory, the interdisciplinary nature of service-learning may preclude a singular methodology. Nevertheless, to not grapple with issues of methodology is to succumb to a default position of again presuming a (false) transparency of experience and practice.

Third, all minors and majors required a field-based experience. This may be an obvious aspect of service-learning programs; what is not obvious, though, is that many of the structured programs constructed sustained, deeply engaged, and consequential experiences. Students in these programs had to spend multiple semesters with a community organization, develop a project that was developed (at least in part) by the community organization, and construct a final project that was public either through its presentation or implementation. Finally, structured academic programs required some form of capstone academic experience. Whether this was a senior seminar or independent thesis, students were required to reflect on and synthesize academic coursework and field-based experiences that spanned multiple semesters within the context of a culminating experience to the overall program's goals.

The structure of a service-learning program matters. But even such programs, at present, have minimal coherence above and beyond the particularities of their specific academic and institutional context. As noted earlier, there was a wide variety of ways by which different programs appropriated and made use of notions of service, democracy, and community. This finding serves as an empirical confirmation of my theoretical point in the last chapter: namely, that there is a gaping hole and thus opportunity to develop an academic grounding by which to articulate a common core of theoretical frameworks, methodological orientations, and empirical case studies for the service-learning field. This is an opportunity to move from toolkits to handbooks, which impacts on and has implications for how service-learning scholars and practitioners talk about and thus organize themselves, their field, and their body of core knowledge.

I want to conclude this chapter by noting that such implications—better structuring service-learning programs and more clearly and coherently articulating the theoretical common core for such a structuring—are not abstract ideals. As this chapter has clearly shown, there are currently numerous scholars in numerous fields already at work in structuring and defining their particular programs and models. The larger issue for the service-learning field is that such practices are occurring haphazardly and without formal and deliberative dialogue across scholars, programs, and organizations. Put otherwise, the service-learning field is already becoming disciplined; the problem is that it is occurring in a helter-skelter and undisciplined fashion. It may thus be time to bring more scholarly attention and consideration to this phenomenon to better understand it and guide it.

CHAPTER 6

THE FUTURES OF SERVICE-LEARNING?

In the previous two chapters I argued that while there was a theoretical opportunity and need for the disciplining of service-learning, the actual on-the-ground realities of most service-learning programs were far from achieving such a goal. In this chapter I present multiple models for potential futures of service-learning as a deeply embedded component within higher education. I suggest that there are in fact multiple means by which to develop sustained and deeply engaging practices in higher education that foster engaged teaching, learning, and scholarship and the role and voice of the community as central components of their programs. These models, I suggest, offer multiple visions of how service-learning may be rethought in the academy. Yet I also conclude that all of these models are premised on and structured around a departmental and disciplinary notion of teaching, learning, and scholarship.

I begin by revisiting the theoretical and empirical limits of contemporary attempts at the institutionalization of service-learning to reemphasize the potential for a disciplinary approach. I then present an extended set of disciplinary examples, focusing on the field of community studies as a comparable and analogous model for service-learning. These examples serve to provide examples of transformation and institutionalization that do not succumb to the worries of marginalizing service-learning as an academic discipline. I also present a countercase—of a discipline (sociology) and its attempt to become more "public"—as an example of the problems of potentially "delegitimizing" an academic discipline. I conclude this chapter with examples of programs that have gone through such a process to demonstrate the potential fruition of such visions.

INSTITUTIONALIZING AND LEGITIMATING AN ACADEMIC FIELD

As I have argued, there are specific theoretical, pedagogical, political, and institutional limits to the current model of institutionalizing a powerful and coherent form of service-learning. To recap, the theoretical limits to service-learning in higher education revolve around tensions between viewing service-learning as a training ground and incubator for the social and civic mission of a public democracy versus the traditional notion of the academic enterprise as concerned primarily with the rigorous, objective, and pure examination of the truth. The pedagogical limits to service-learning in higher education refer to the reality that very few students meet the traditional notions of what constitutes the "servers" while, at the same time, the faculty engaged in service-learning practices are not centrally located, nor are they the norm in the academy.

The political limits to service-learning reside in the fact that service-learning has a progressive and liberal agenda under the guise of a universalistic practice that fosters an unjustified valorization of the goals of civic engagement and presumes social justice to be based upon a teleological upward movement from charity-based forms of volunteerism toward justice-oriented modes of sustained and collective practice. Finally, the institutional limits to service-learning reside in the realization that higher education works by very specific disciplinary rules about knowledge production, people who have the academic legitimacy to produce such knowledge, and the manner in which they can do so.

The limits just outlined are fundamentally linked to the undergirding of theoretical presuppositions of contemporary service-learning theory and practice. Namely, service-learning is presumed to be a politics by which to transform higher education. As such, service-learning becomes positioned within the binary of an "oppositional social movement" embedded within the "status quo" academy. Moreover, this perspective reifies (and thus presumes) service-learning to be a coherent and cohesive pedagogical strategy able to see its own blind spots as it pursues liberal and always liberatory agendas.

Yet as I have argued throughout this book, such presumptions are theoretically and empirically unfounded. Thinking about service-learning as a politics to transform higher education is a theoretical cul-de-sac that, by reifying the service-learning experience, undermines any analytic opportunity to understand how and why it is ultimately deeply limited. The possibilities for service-learning, I thus suggest, lie in embracing rather than rejecting the very academy the service-learning movement is attempting to transform. More precisely, it is to speak about service-learning as akin to an academic discipline that has the ability to control its knowledge production functions by internally debating and determining what issues are worthy of study, by what modes of inquiry, and to what ends. This presumes a plurality of perspectives on what service-learning is and should be. It presumes that the scholarship surrounding service-learning is not solely centripetal or convergent in focus.

In fact, constructing a discipline-based, academically governed, and bureaucratically stable "academic home" for service-learning may be the most productive means by which to spread the scholarship of engagement across the academy. For, as I suggested earlier, Wendy Brown's (2003) critique of women's studies ultimately does not engage the (lack of a) future of women's studies; rather, it engages the inadequacy of viewing women's studies as the revolutionary vehicle for a feminist liberation.

Revolutions, Brown argues, presume a coherency and liberatory status that women's studies never had (see Moraga and Anzaldua 1981, for just such a critique of "first wave" feminism). For Brown (2003), such a throwing off of the yoke of liberation is itself liberatory: "If we are without revolutionary possibility today, we are also free of revolution as the paradigm of transformation" (15). Women's studies as an academic discipline thus has the freedom—in fact the obligation—to develop, question, and revise its own tools, its own practices, its own analytic foci, and its own disciplinary modes of knowledge production and dissemination. This suggests that only by becoming disciplined can service-learning truly be sustained and nourished in the academy. Such a disciplinary move of achieving programmatic status, I suggest, is a crucial component for scholars across the academy to institutionalize and insulate their areas of study as legitimate and worthy undertakings.

At the heart of this argument is the notion that the disciplining of a movement is a necessary precondition for its ability to work within and through the context-specific mechanisms of higher education. While higher education is certainly buffeted by external pressures (as the final chapter of this book details), the spread of counternormative paradigms occurs primarily sub rosa. As Rojas (2007) argues of black studies:

> Black studies' institutionalization shows that movements test cultural boundaries; they do not mimic them, but expand them through hybridization. Social movements expand the "institutional vocabulary" of a field such as higher education by questioning what is acceptable and extracting compromises between current behavioral norms and the movement's demands. Thus, the construction of black studies was not guided only by "institutional logics" that enforce conformity within higher education. Rather, the black studies movement generated a range of alternatives, some of which were modified so they could be deemed acceptable to at least a few university leaders. The cultural imperatives of higher education were used to discard proposals that were too radical, but that left many proposals that subtly changed the criteria of acceptable academic work. Thus, if movement activists can gain a sufficiently strong understanding of bureaucratic processes and outcomes, they can alter the organization's logic. (214–15)

It is thus possible to view the disciplining and departmentalization of a field as accommodating to and/or modifying the notion of the role of an academic discipline. Recent examples (I could have chosen many others) include the rise of Jewish studies and international studies that attempt to mirror and replicate the institutionalizing moves of women's studies and black studies in the 1970s. Jewish studies, for example, arose in the post–World War II era and has recently developed into a small yet thriving (inter)disciplinary organization. The field has its own umbrella organization—the Association for Jewish Studies (AJS)—that supports almost 2,000 members, an annual meeting, a scholarly journal, and a central hub for resources such as funding and a catalog of courses. Likewise, the international studies field is organized through an umbrella organization—the International Studies Association—that coordinates dozens of affiliated suborganizations, supports an annual conference, and sponsors five academic journals in the field.

Both fields, it should be noted, much like women's studies, continue to debate and develop their own internal notions of identity and positioning in the academy. Scholars within Jewish studies have had ongoing discussions at a host of venues (e.g., AJS 2006) that engaged key questions of who their clientele was (Jews versus non-Jews) and how this impacted the curriculum taught; the interdisciplinary nature of the program and the concomitant pros and cons, including access to a wide range of scholars and dependence on nondepartmental faculty and budgets; and whether a common core existed in the major, and whether this constituted the historical, language, literary, or other cultural focus. As one commentator noted (Hyman 2006):

> Jewish studies is a form of cultural or area studies, more akin to American studies or religious studies, which presumes that multiple methodologies are necessary for their study, than to traditional fields like history or sociology. The recognition of the interdisciplinary nature of Jewish studies is quite appropriate and promotes further research, but it also raises the question of what enables students in Jewish studies programs or departments, with their diversified courses, to feel that they participate in a common field. (22)

Likewise, the international studies (IS) field, a subset of political studies and akin to international relations, has hundreds of programs across postsecondary institutions; yet it is only in the last five years that scholars in the field have begun to systematically examine exactly how the field is structured and the implications of such a structuring (e.g., Brown et al. 2006; Ishiyama and Breuning 2004a, 2004b; Hey 2004). Ishiyama and Breuning (2004a), for example, note that: "Despite the increasing popularity of IS majors, there appears to be very little consensus regarding their features. Some are only loosely structured, affording wide latitude to students in terms of course selection. Other programs are much more structured, emphasizing a set of key common courses that act as a core of the international studies curriculum" (134). Ishiyama and Breuning (2004a), following AAC&U (Wahlke 1991) recommendations, argue that "only a consciously structured major that entails sequential learning promotes the development of the 'building blocks of knowledge that lead to more sophisticated understanding and . . . leaps of the imagination and efforts at synthesis'" (136–37). Cumulative knowledge built up by more and more electives, they argue, do not move beyond shallow learning.

These brief examples suggest that the internal self-reflexive questioning of a field and its intellectual and physical boundaries do not in any way undermine its academic integrity and legitimacy. Quite the opposite; such critical engagement is viewed as a key feature of the sustainability and adaptability of a field to disparate and changing conditions across higher education. Yet a counterexample of the reversal of such a process—the attempted move from internal deliberation to the construction of public relevance—may have less sanguine consequences.

The field of sociology has embraced the notion of a "public sociology" movement with decidedly mixed results. The term "public sociology" itself was introduced by Herbert Gans (1988) when he argued that sociologists should more strongly align their work to inform public issues and debate. Yet it was Michael Burawoy's (2004a) presidential address to the American Sociological Association that set the current stage of examination and debate within the field. Burawoy (2004a) argued that public sociology "defines, promotes and informs public debate about class and racial inequities, new gender regimes, environmental degradation, multiculturalism, technological revolutions, market fundamentalism, and state and non-state violence." Moreover, Burawoy (2004b) suggested that public sociology was but one of four typological norms of sociology that fit within a two-by-two matrix of the division of sociological labor.

Burawoy's (2004b) typology differentiated between instrumental and reflexive knowledge and between academic and extra-academic audiences. He suggested that traditional sociology (which he termed "professional sociology") focused on instrumental knowledge for an academic audience while public sociology focused on reflexive knowledge focused on extra-academic audiences. Filling in the matrix, "critical sociology" focused on reflexive knowledge for academic audiences, whereas "policy sociology" focused on instrumental knowledge for extra-academic audiences. Each type of sociology, Burawoy argues, informs and is in interdependence with each other. And, interestingly, such a typology, he argues is applicable to every field.

While such a stance has been roundly embraced by many sociologists, many have been skeptical and deeply disturbed by such a public focus. As Title (2004) summarizes, public sociology "assumes an unjustified moral superiority. It jeopardizes accomplishment of goals that would make sociology genuinely useful. It is somewhat dishonest in claiming more than can be delivered and in the process undermines sociological credibility. And it often patronizes those outside the profession" (1643). The idea that sociologists have definitive answers that can be simply and transparently relayed to the public is problematic from such a critical view, exactly because it ignores sociology's own arguments regarding the problematics of neutrality on complex matters, the inability to rely solely on disciplinary expertise, and the false singularity of voice for a highly diverse field.

Other sociologists have been even more caustic. Deflem (2007), for example, has argued that public sociology is neither public nor sociology. It is not public in that "it is a particularistic political position" that speaks solely from its predefined socialist Marxist perspective. Such a politicization in turn threatens the very standing of sociology upon which public sociologists have claimed their strength of offering to the public—their disciplinary expertise of examining complex societal issues. Instead, Deflem (2007) argues: "A relevant sociology requires an unyielding commitment to sociology as an academic discipline. The courage to be resolutely analytical about society is the true revolutionary quality of sociology. In its inability to be resolutely analytical and transcend the blinding darkness of fundamentalism, public sociology is tame at best, conservative at worst." Title (2004) offers an empirical example of the difficulty of speaking outside disciplinary bounds:

> Advocates [of public sociology] seem to think that what is 'socially just' is clear and easily agreed upon among people with good will or sociological training. Actually, almost every social issue involves moral dilemmas, not moral clarity. What is or is not 'just' is almost never unambiguous . . . In one of his writings Burawoy lists the spread of disease as one of the goals of a just society and therefore one to be pursued by "public sociologists." On a superficial level, most people would readily agree. Preventing diseases, however, often involves restrictions on human freedom and hard decisions about allocation of scarce products or services . . . Do sociologists have any way of knowing the proper tradeoff of sacrificing X amount of freedom for saving Y number of lives? . . . To assume that we do have such superiority [of policy knowledge] and to expect people to accord us respect on that basis is really quite arrogant. (1640)

The worry for these sociologists is that scholarship attuned to and focused on "extra-academic" audiences (which may be akin to contemporary notions of public scholarship [e.g., Cohen 2006; Colbeck and Wharton-Michael 2006; Wharton-Michael et al. 2006]) undermines and thus delegitimizes the very academic enterprise of scholarship that allows such scholars a voice in the first place. This is not to suggest that scholars cannot speak to different audiences in different ways; nor does it mean that one cannot have at the same time scholars with different foci and modes of scholarship. Rather, what is at stake is the presumption that so-called instrumental knowledge production for academic audiences serves primarily and fundamentally as a staging ground for more public knowledge for public audiences with policy implications.

Jewish studies and international studies (and, for that matter, women's studies and physics) all of course have public consequences.

Academics in all of these fields speak out to some extent or another when their expertise can help clarify or inform public issues and developments. But such "public scholarship" may be better viewed as a by-product of the academic enterprise rather than as its primary and sole goal. Most academic developments and findings rarely see the light of day. While this may of course bury some legitimate and important findings, more commonly this is because the academy, and the peer review process upon which it is built, works. It works just fine because other scholars question the findings, extend or modify its parameters or target populations, question and revise its assumptions, and rework its conclusions.

Such a deliberate and deliberative process does not align well with the public purpose of public scholarship. Tentative findings based on shaky presuppositions that may be undermined or modified in the next journal issue do not work well as guiding manifestos for community change or urban revitalization. And for scholars to attempt to wring more out of their findings than can be genuinely done only serves to disenchant a public that may feel bamboozled (yet again) by "authorities" claiming more than they can deliver.

The question that thus remains for the service-learning movement and public scholarship more generally is how academic processes can work through and toward public sensibilities. Or to put it into a more binary phrasing: rather than attempting to make the academy focused outward toward the public, how can we bring public issues into the academy? How can community issues become academic?

I suggest that such examples can indeed be found across the academy. They may be historically dated or haphazardly scattered in diverse institutions and departments; but there are indeed useful precedents for linking community issues and community engagement with deeply academic goals and practices. The rest of this chapter thus explicates the case of community studies (a subfield of sociology), and offers several contemporary examples of departments that have genuinely integrated public questions into the academic enterprise.

SERVICE-LEARNING AS COMMUNITY STUDIES?

Community studies was once a thriving subfield of sociology during the 1950s and '60s and is today experiencing a revival of sorts across the humanities and social sciences (Adams 2000; Crow 2002). The subfield originally went into disrepute by the early 1970s as its conceptions of "community" were deemed theoretically incoherent, its methodology suspect, and its conclusions irrelevant in our "postindustrial" world, where technology, media, and high-speed transportation seemingly have destroyed all notions of the continuity and relevance of place (Bell and Newby 1972; Stacey 1969). Yet today, as higher education attempts to appropriate Boyer's (1990) "scholarship of engagement," it is possible to see how community studies may be able to bridge and combine academic rigor with community engagement.

To understand what community studies can offer, though, we must first be able to answer "what is community studies?" and, prior to that, "what is community?" This is far from simple. In a 1955 paper (Hillery 1955), in what has come to be seen as the *piece de resistance* of the field in its heyday, George Hillery surveyed ninety-four different definitions of community and found that the only point of commonality was that "all of the definitions deal with people. Beyond this common basis, there is no agreement" (20).

While probably hyperbolic, Hillery's point is well-taken: there is no definitive definition of community. Most early definitions used Ferdinand Tonnies' (1887/1957) distinction between *Gemeinschaft* and *Gesellschaft*, which are usually translated as "community" and "society." *Gemeinschaft* is the local, the personal, the intimate, moral certainty. *Gesellschaft*, on the other hand, is the opposite: the large scale, impersonal, economically based, and utility-driven marketplace with no foundational sense of virtue, morality, or home.

Bell and Newby (1971) suggest that "for Tonnies, there are three central aspects of Gemeinschaft: blood, place (land), and mind, with their sociological consequents of kinship, neighbourhood and friendship" (25). This trinity of familial bonds, of a bounded geographic locale, and of interest-based relationships is the skeleton key for understanding how communities have been traditionally viewed and studied.

From this perspective, then, community studies was nothing more or less than the analysis of individuals' behavior (and the community's impact on such behavior) within these three intertwined spheres. The Lynds' *Middletown*, William Whyte's *Street Corner Society*, and the "Yankee City" project led by W. Lloyd Warner are the emblematic examples here. One may also think of the polar opposite—the studies of *Gesellschaft*, such as the classics Whyte's *Organization Man* or David Reisman's *The Lonely Crowd*.

But as I mentioned earlier, such notions of "community" fell into disrepute by the late 1960s and early 1970s. The federal highway system, housing patterns, and affordable transportation all seemingly collapsed the notion of distance and thus of place; mass media and technology permitted easy communication at great distances, thus loosening the bonds of locality even more. All of these contributed to the notion that this thing called "community" may be more mythic than real, more a straw man than an analytic framework for "real" social scientists. It did not help that the methodology employed by many of these studies came to be seen as amateurish, overly subjective, and lacking in analytic rigor. "The poor sociologist's substitute for the novel" is how one critic phrased it (Glass 1966, 148, quoted in Bell and Newby, 13).

So with no specific and bounded community, community studies fractured. Today, scholars across the humanities and social scientists continue to examine community-based issues through diverse methodological and disciplinary lenses such as network studies, locality studies, community health, urban studies, or community development. Scholars are now finding "communities" online, within a single block in an urban metropolis, or amongst drop-in clients to a specialty store. While still making use of the tripartite division, such scholars have also sharpened the analytic frame of what constitutes a community: (1) a locality; (2) a "local social system" that refers to a set of social relationships or networks; (3) a type of relationship (Crow 2000, 2002; Freeland 2004; Stacey 1969; Vasta 2000). Additionally, there are multiple means to frame how one looks at any specific community: communities may be framed as ecological/biological functioning mechanisms; as organizations; as microcosms; and as types (the rural-urban[-metropolis]).

Given this diversity of what constitutes a "community," I suggest that there are three specific ways to conceptualize community studies: community studies as (a) a methodology (akin to ethnography or community-based research); (b) a pedagogical practice (akin to service-learning); and (c) an analytic lens (akin to American studies or women's studies).

Community studies as a methodology is a mode of examining an issue, be it public health or poverty. Community studies in this perspective is one tool in the toolkit of academic scholars engaged in research or analysis. Thus much like ethnography, statistical analysis, or survey research, community studies is a holistic investigation of how an issue plays out within a community (however defined). A quick search on any academic database turns up numerous studies within the medical and health care fields that take such an approach. Community studies, for these scholars, is a specific empirical procedure that focuses on a specific, bounded, and holistic locale.

Community studies as a pedagogical practice is a model of teaching, one that links postsecondary students with their local and global communities. As one community studies center (Naropa 2007) phrases it, "Community Studies means studying and working with communities." This may be thought of as another form of service-learning whereby college students gain insight into their studies through real-world and real-time experiences.

Finally, community studies as an analytic lens is a conceptual framework by which to examine an issue. An analogous example would be women's studies. Thus the community is not simply the object under analysis, but the frame within which one studies an issue. Analogously, women's studies scholars would suggest that "women" are not solely the objects studied; rather, feminist perspectives inform the analysis of specific issues. From this perspective, one is not simply a historian or sociologist investigating a community-based issue; rather, one is a community studies scholar investigating an issue with historical or sociological import.

This issue can now be explored inductively by examining what community studies actually looks like on the ground in higher education. The guiding question is how do academic units incorporate the study of the community into their daily practices. Table 6.1 provides an overview.

An analysis of these twenty-one programs' self-description (based on their Web sites) revealed three distinctive "community studies"

Table 6.1 Community studies academic programs in higher education

Concentrations
- California State University – Northridge's Asian American Studies department offers a community studies concentration
- Clemson's Sociology department offers a community studies concentration
- George Mason University's New Century College offers a community studies concentration
- Guilford College offers a community studies concentration within their justice and policy studies major
- Portland State University offers a community studies cluster within their University Studies program
- University of Missouri-Columbia's Department of Rural Sociology has a community studies emphasis

Minor
- Ferris State University offers a community studies minor
- Santa Clara University offers a community studies minor
- University of Michigan's College of Literature, Science, and the Arts offers an urban and community studies minor
- Washington State University's Community & Rural Sociology Department offers a community studies minor

(*Continued*)

Table 6.1 Continued

Major
- University of Baltimore offers a community studies and civic engagement major
- University of Massachusetts - Boston's College of Public & Community Service offers a community studies major
- University of Utah's Department of Family & Consumer Studies offers a consumer & community studies major

Graduate offering
- Northeastern University's Department of Sociology & Anthropology offers an urban affairs & community studies concentration for its graduate program
- University of Illinois at Urbana-Champaign's Department of Human and Community Development offers a PhD program in community studies and outreach
- University of Vermont offers a graduate program in education and community studies

Department
- St. Cloud State University has a Community Studies department
- University of California-Santa Cruz has a Community Studies department
- University of Connecticut has an Urban and Community Studies department
- University of Maine-Machias offers a behavioral sciences and community studies major
- University of Wisconsin, Milwaukee has a Department of Educational Policy and Community Studies

models: (1) community studies as methodology; (2) community studies as academic specialization; and (3) community studies as community development and social change. While many programs spoke of community studies as a methodology, such a methodology appeared primarily to be a precondition for accomplishing either (or both) of the latter two goals. I thus focus on these latter two goals. It is also noteworthy that while several programs clearly articulated both the academic specialization and social change goals, most of the programs fit clearly within a single category.

Community studies as academic specialization very clearly views community studies as a means to the better understanding and analysis of a specific issue. As Northeastern University's urban affairs and community studies concentration states: "This concentration focuses on an analysis of various social issues, social problems, and policy responses that are of particular importance to cities and suburbs." Alternatively, *community studies as community development and social change* focuses much more on an activist orientation. As Portland State University's community studies cluster (2007) states:

> This cluster explores the nature of the communities we live in, whether defined spatially (such as a neighborhood) or as a set of ties

> based on sharing a common interest. Building community has become a central debate in a number of social sciences, including sociology, political science, economics, and psychology. In a culture emphasizing individualism and individual rights, how can needs for community and responsibility to others be balanced? *Thus, in this cluster, students have the opportunity to gain practical as well as theoretical experience with building communities.*

It is important to note, though, that more "activist" programs are structured within and through traditional academic course content. Students in the Portland State cluster must take a core course entitled "Community Studies" as well as two other courses from a range of traditional offerings such as "Global Health," "Economics of Race and Ethnicity," and "Introduction to Urban Planning." Thus irrespective of where programs fall on the academic-activist spectrum, all presume that "community studies" is an analytic construct. Put otherwise, all programs, by virtue of functioning within a traditional higher education structure, are able to develop and sustain their own "brand" of community studies irrespective of its positioning as "activist" or "academic."

Such is not the case with research centers. Ten research centers were found through the same search process:

Dickinson College's Community Studies Center
Elon College's Elon Project for Ethnographic Research and Community Studies
Ferris State University's Community Studies Institute
Jefferson Community College's Center for Community Studies
LaGrange College's Center for Community Studies
Loyola Marymount University's Center for the Study of Los Angeles (it has a programmatic focus on community studies)
Naropa University's Community Studies Center
Northeastern University's Center for Urban and Regional Policy
University of Colorado at Boulder's INVST Community Studies
Vanderbilt University's Center for Community Studies

A review of these centers' descriptions and mission statements revealed that the primary focus was either on methodology or pedagogy. Methodologically, such centers either promoted a decidedly ethnographic focus (e.g., Dickinson, Elon) or a policy focus on local and regional community issues (e.g., LaGrange, Vanderbilt). Dickinson's Community Studies Center, for example, states that "the Community

Studies Center at Dickinson was developed 5 years ago as an effort to support and extend faculty-student fieldwork, including ethnography, participant-observation and oral history." Vanderbilt's Center, alternatively, focuses on "research, dissemination, outreach and capacity-building components." Several of the centers (e.g., Naropa, University of Colorado) focused on community studies as a pedagogical practice to support students' engagement with local communities and community issues. The University of Colorado-Boulder's INVST mission statement, for example, states that, "Our programs develop engaged citizens and leaders who work for the benefit of humanity and the environment."

A fascinating and important distinction thus becomes quickly apparent: while there are about two dozen academic programs and a dozen research centers scattered haphazardly around the country all employing the "community studies" moniker, all of the academic programs view this as an academic enterprise (along a spectrum of scholarship-activism), while all of the research centers focus on community studies as a methodological or pedagogical practice. I am sure that there may be counterexamples of research institutes that do not conform to this model. Yet the findings are highly suggestive—and mirror the findings in the previous chapter on service-learning minors and majors—that it may only be possible to construct legitimate and long-standing institutional programs within the departmental structure.

What community studies programs have done (whether as concentrations, minors, or majors) is to take the community as its locus of investigation and institutionalize such practices. I have already shown the possibility of such a structuring specifically focused on service-learning in the previous chapter. The examination of Community Studies programs demonstrates that such practices of institutionalizing engaged scholarship is, if not a commonplace, a legitimate model in higher education.

WHAT FUTURES FOR SERVICE-LEARNING?

Ultimately, the second part of this book has argued that the service-learning movement may have the wrong unit of analysis as its fundamental agent of change. Strengthening the scholarship of engagement and service-learning at both the institutional and individual levels is of course important. One needs students and faculty to buy into the practice; and one needs institutional leaders' commitment. Yet it is this "middle ground" of departmental policies and practices that appears to truly matter for the sustained and consequential engagement with and impact on a community.

It is at the departmental level that pedagogical practices and disciplinary norms get debated and settled; it is at the departmental level that tenure and promotion assumptions and standards get set and promulgated; and it is at the departmental level that students' and faculty's identities and perspectives become forged and focused. If service-learning is to have a productive future in the academy, it may need to investigate a range of multiple futures by which it becomes implemented and sustained.

To leave it at the departmental level is of course in one respect a gamble with the direction in which such practices may go. As the empirical examples in the last two chapters demonstrate, some departments may focus on activist orientations that manifestly incorporate community voices and individuals, while others involve themselves in formal academic and scholarly examinations of the notion of community. Some departments may focus their service exclusively on a specific and local geographic locale, while others send the majority of their students across the country and overseas for their service experiences.

The diversity and plurality of what constitutes a scholarship of engagement thus becomes refracted through the particularities of communities, institutions, and individuals. In turn, it suggests that the sustenance of a transformative and powerful version of service-learning may reside as much in the details of what we do and debate as scholars as in a broad social movement with a seemingly cohesive vision. The work of institutionalizing community engagement must occur in the day-to-day practices of higher education, and it is this notion of faculty work and how it can be done that the last section of this book takes up.

Part III

Embracing a Scholarship of Engagement in Higher Education

CHAPTER 7

TOWARD A THEORY AND PRACTICE OF COMMUNITY ENGAGEMENT

Public scholarship, community-based research, undergraduate research, the scholarship of teaching and learning, community engagement, project-based learning, experiential education, cooperative education, civic engagement, peer review of teaching, deep learning, problem-based learning, cognitive apprenticeship, integrative learning, participatory research, the list goes on and on. "The list," that is, of complementary, analogous, and related branches of service-learning, and, more broadly, of community engagement. For service-learning is one amongst multiple examples of the academy's response to internal and external pressures for relevancy. Such pressures can be traced to the 1980s outcry against the disengaged academic (e.g., Lucas 1996; Sommer 1995), a genuine and long-overdue return to the civic mission of higher education (e.g., Harkavy et al. 2007), or to faculty's embrace of a public scholarship and lost public intellectualism (e.g., Burawoy 2004a).

Irrespective of the precise history, the notion of community engagement as a form of a "scholarship of engagement" has become an accepted part of the nomenclature within higher education. And in one respect, service-learning fits snugly under this umbrella in that it is exactly the linkage of engagement in the community with scholarly practices and research. Boyer's (1990) fourfold typology clearly delineates a scholarship of application whereby we build bridges between scholarly knowledge and community needs.

But I want to suggest in this last section of the book that service-learning scholars and practitioners may need to reexamine and rethink what they do and how they do it. Namely, I want to suggest that we attempt to view service-learning and community engagement as simply everyday aspects of our jobs as faculty and administrators in higher education. Whether we do this in our individual courses, within departments focused on civic engagement, or through an entire institution's mission and vision, a scholarship of engagement should be no more novel or noteworthy than any other scholarly task.

I take this point from Stanley Fish's (2003, 2008) provocative intonation that we in higher education "save the world on our own time." Ever the contrarian, Fish (2008) argues that higher education should do no more (and no less) than focus on its main purpose of producing, questioning, and disseminating knowledge within its disciplinary specializations. To go any further—to strengthen students' civic and political engagement, foster diversity, or end racism—is foolhardy and destructive of the role and function of higher education in our society. Fish appears out of alignment with and going against the grain of current sentiment in higher education, as community engagement becomes ever more prevalent and embraced within higher education (as the list of monikers cited above demonstrates). Yet I want to argue in these final chapters that not only is Fish right but that the practice and theory of community engagement can only benefit and prosper if faculty and administrators in higher education understand and embrace Fish's seemingly contrarian arguments.

This chapter provides a means for faculty to begin to view how community engagement is in fact another means of being good scholars—however defined within their respective fields—devoted to the essential practices of higher education: the production and dissemination of knowledge and search for truth. I use the fourfold typology articulated in the first chapter as my grounding to suggest the next level for deepening and strengthening faculty's engagement with community-based models of teaching, learning, and research.

Moreover, I suggest that community engagement is in fact the most productive theoretical conclusion to my arguments in the first part of this book that one cannot teach "without remainder." Put otherwise, I suggest in this chapter that a rethinking of service-learning as a model of engaged scholarship fosters a deep linkage and synthesis of the theoretical insights of the first part of this book with the empirical and structural findings in the second part.

In the concluding chapter, I place service-learning and community engagement within the larger picture of the major contemporary trends and tensions within higher education. I suggest that due to three major issues—demographic patterns, market pressures, and the nature of faculty work—the service-learning movement, as we have known it for the last twenty years, may not be as stable or secure as we may think. I thus outline how the reframing I begin in this chapter—toward embracing community engagement—may be one of the few (and best) means to sustain the power and transformational potential of community-based models of teaching, learning, and research. But to get there, I need to first back up and explain why Stanley Fish is right that we should "save the world on our own time."

STANLEY FISH, COMMUNITY ENGAGEMENT, AND SAVING THE WORLD

In many ways, Fish's (2008) recent book is the public culmination of an academic endeavor begun more than thirty years ago in his groundbreaking analysis of Milton specifically and literary theory more generally (Fish 1972). Literature such as Milton's, Fish argued, function as "self-consuming texts" that "do not allow a reader the security of his normal patterns of thought and belief" (409). The point of such texts is exactly to avoid having the reader come to a final point: "Coming to the [final] point fulfills a need that most literature deliberately frustrates (if we open ourselves to it), the need to simplify and close" (410; see Butin 2005b, for a deeper explication of this argument).

This antifoundationalism of resisting and rejecting all "objective" and thus static readings of a text is also what is at the heart of how Fish has come to view the role and function of the academy: as being constantly engaged with and in pursuit of a never closed-off truth. For Fish, truth is always temporary and always being reformulated within the context of our particular "interpretative communities" (Fish 1982), which, in higher education, means our particular disciplinary academic communities. As such, the only way we can do our jobs is to have the authority, the academic legitimacy, to control the means and goals of our own knowledge production and dissemination.

Fish is a master at phrasing complex ideas simply and concisely. "Do your job," says Fish, "don't do somebody else's job and don't let someone else do your job" (8). According to Fish, colleges and universities have vastly and inappropriately overreached in what they claim to do and be for the students who attend them and the communities within which they reside. Higher education, claims Fish, should not be in the business of enhancing or expanding students' moral, civic, or social characters; nor to inveigh on current social, cultural, and political issues such as war, poverty, or racism; nor to revitalize, transform, or collaborate with local and regional communities.

Rather, Fish suggests, "College and university teachers can (legitimately) do two things: (1) introduce students to bodies of knowledge and traditions of inquiry that had not previously been part of their experience; and (2) equip those same students with the analytical skills—of argument, statistical modeling, laboratory procedure—that will enable them to move confidently within those traditions and to engage in independent research after a course is over" (13).

In his words, academics should "academicize" in their pedagogy. This is a highly focused (and some might suggest myopic) vision for higher education, and Fish is very well aware of this. "The topics considered and arguments waged in these chapters vary," Fish writes, "but everything follows from the wish to define academic work precisely and narrowly" (17). Higher education, and the faculty teaching and researching within it, has a very specific function: knowledge production and dissemination. There are, of course, intended and unintended byproducts to these actions; and there are, of course, cocurricular and extracurricular functions to which higher education increasingly attends to; and the university is, of course, a part of a larger web of social, cultural, economic, and political interconnections. But ultimately, for Fish, "the pursuit of truth is the cardinal value of the academy" (119). And this pursuit of truth, wherever it may lead, without end, without undue external encumbrance, done diligently and carefully within the bounds of academic practice, is what drives Fish's arguments.

In one respect, Fish draws out this argument to its fairly technocratic and pragmatic conclusion: doing one's job as a faculty member is hard enough already. Do not, as such, pile up other tangential responsibilities, no matter how high-sounding or politically expedient. Faculty are "responsible for the selection of texts, the preparation of a syllabus, the sequence of assignments and exams, the framing and grading of a term paper, and so on" (57). Their goal is to help students master specific bodies of knowledge and particular scientific procedures, to understand certain worldviews and their limits, to come to think and view specific issues in the world from particular disciplinary lenses (be it physics, history, or engineering) in which they were prepared. To go beyond that—to, in his words, not stick to one's job—undermines one's contractual agreement to be a good pedagogue. Fish does not mince words here: "Responsibility of a pedagogical kind seems to exist in an inverse relationship to noble aspirations in the education world" (2005, 43).

This is over-the-top and standard-fare Fish. For Fish is deeply concerned about the power of pedagogy. Fish wants, implores, demands, that faculty teach, and teach well. By this Fish means that our job as academics is to "academicize" (i.e., read: not proselytize): "To academicize a topic is to detach it from the context of its real world urgency, where there is a vote to be taken or an agenda to be embraced, and insert it into a context of academic urgency, where there is an account to be offered or an analysis to be performed" (27). For Fish it is irrelevant whether I am a Democrat or a Republican, gay or straight, a Red Sox or Yankees fan. When I walk into a classroom all that matters is the pursuit of truth, as defined by academic standards, protocols, histories, and lines of inquiry.

Fish's point that we should save the world on our time is thus but another way of reminding us that the academy is a highly specialized institution with highly specialized functions. It cannot be and do everything for everyone. "Distinctiveness," says Fish, "is a prerequisite both of our survival and of our flourishing. Without it we haven't got a prayer" (100). For if we do not attend to what we do best—which is developing, testing, critiquing, transforming, and transmitting knowledge—we open ourselves up to the critique of dealing with issues for which we are not equipped or trained. This does not mean that the academic space is thus somehow sterile, pure, or neutral: "Again, this is not to say that academic work touches on none of the issues central to politics, ethics, civics, and economics; it is just that when those issues arise in an academic context, they should be discussed in academic terms; that is, they should be objects of analysis, comparison, historical placement, etc." (25).

What Fish has done here (brilliantly, in my perspective) is to attempt to safeguard the mission and practice of higher education by placing the criteria of success internal to the workings of the mission and practice of higher education. My status, success, and legitimacy as an academic are not just convergent with what and how I teach and research; what and how I teach and research are who I am as an academic. I am thus not bound by political pressure, students' desire for "balance," or even Stanley Fish's prognostications about how I should teach my course. I am simply and solely bound to the pursuit of truth as I best see fit within my own academic and interpretive community.

It is here, I suggest, that Fish's arguments cut directly to the heart and soul of how many in the community engagement movement see themselves: "The view I am offering of higher education is properly called deflationary; it takes the air out of some inflated balloons.

It denies to teaching the moral or philosophical pretensions that lead practitioners to envision themselves as agents of change or as designers of a 'transformative experience,' a phrase I intensely dislike . . . Teaching is a job, and what it requires is not a superior sensibility or a purity of heart and intention . . . but mastery of a craft" (53). This "mastery of a craft" is, for Fish, what will save higher education because it will allow us as faculty to focus on our legitimate jobs of academicizing any and all issues; such academicizing leaves behind cultural transformation and partisan politics in favor of the search—in the classroom and in one's scholarship—for the always complex and contingent truth.

Fish sees his perspective as antithetical to what is commonly thought of as the community engagement movement; I, though, see it as a perfect roadmap for the legitimation of certain parts of this movement in higher education. The linkage begins here: "What set me off in all of this," Fish comments in a 2005 interview about his essays about the politicized classroom (i.e., Fish 2003a, 2003b), "was the book called *Educating Citizens: Preparing America's Undergraduates for Lives of Moral and Civic Responsibility*" (44). This book (Colby, Ehrlich, Beaumont, and Stephens 2003) and a subsequent book (Colby, Beaumont, Ehrlich, and Corngold 2007) from the Carnegie Foundation for the Advancement of Teaching provide some of the theoretical grounding and empirical ballast for the value of community and civic engagement in higher education from one of the two or three most prominent organizations in the higher education field.

Fish's attack on this book was that such noble goals, while laudatory in a participatory democracy, are not only inapplicable in higher education, they are in fact harmful and corrosive to the mission and practices of higher education. "Aim low," he suggests to the reader.

> You can reasonably set out to put your students in possession of a set of materials and equip them with a set of skills (interpretive, computational, laboratory, archival), and even perhaps (although this one is really iffy) instill in them the same love of the subject that inspires your pedagogical efforts . . . You have little chance however (and that entirely a matter of serendipity) of determining what they will make of what you have offered them once the room is unlocked for the last time and they escape first into the space of someone else's obsession and then into the space of the wide wide world. And you have no chance at all (short of a discipleship that is itself suspect and dangerous), of determining what their behavior and values will be in those aspects of their lives that are not, in the strict sense of the word, academic. You

might just make them into good researchers. You can't make them into good people, and you shouldn't try. (2008, 58–59)

At one level Fish is just pointing out what developmental researchers (e.g., Allport 1954; Baxter Magolda 1999; Tatum 1992; see also Butin 2005c) have consistently noted: that changing one's point of view (especially on a contested or controversial topic) takes many years and much investment of time and energy on the part of an institution, the faculty, and the students themselves. On another level, though, Fish wants to cleanly and clearly demarcate where the job of higher education begins and where it ends.

Specifically, Fish is demanding that knowledge production and dissemination begin and end in higher education, be it in the research lab or in the college classroom. This is what we have control over—nothing less and nothing more. Put otherwise, the legitimacy of what we do as academics can only be determined by the internal conditions we ourselves have set for it. We control our own knowledge legitimation. Fish's attack on Colby et al. (2003) is thus a demand that we as researchers and professors stick to what we know and do best; and if we do, we will have a much better chance of being good teachers who positively impact our students, good researchers who can inform and extend our disciplinary knowledge, and most important of all, good and politically-savvy academicians who can actually stay in control of how we define and go about our jobs as good teachers and good researchers. What Fish saw in *Educating Citizens* was his greatest fear realized: prominent academics (supported by a prominent organization) seemingly claiming that higher education could do and be all things to everyone.

This brings me back to my claim of the value of Fish's arguments for community engagement. What I want to suggest is that Fish's critique is in fact deeply compatible with what I define as the technical and antifoundational aspects, and not with the cultural or political. Specifically, when community engagement (or any form of pedagogy, for that matter) attempts to directly enhance diversity or promote social justice or strengthen civic virtues, the first questions become: Whose notion of diversity? Whose notion of justice? Whose notion of virtues? These are morally and culturally fraught issues that demand allegiances and commitments and, by definition, invoke partisan and oftentimes binary distinctions. But for Fish, the university must be amoral: "the university gives no counsel, and that is the professional, and in some sense moral, obligation of faculty members to check their moral commitments at the door" (101). Our job is to academicize the critical issues of diversity, justice, and civic virtues, not preach about them.

This is why community engagement from technical and antifoundational perspectives can be truly powerful, sustainable, and defendable as a deeply academic practice. As a technical practice, it is about helping us do our job—as academics who academicize—better. Whether I am teaching about income inequality, math education, or nonprofit management, community-based pedagogical practices can help students understand the contextual realities, real-world subtleties, and multiplicity of perspectives of the specific issue under analysis. Fish, in fact, acknowledges and praises such active engagement (even as he does not understand service-learning, and thus conflates it with community service): "a student who returns from an internship experience and writes an academic paper . . . analyzing and generalizing on her experience, should get credit for it" (21). This is, for Fish, higher education at its best: introducing students to bodies of knowledge and ways of thinking that help students academicize deeply volatile and complex issues through careful reflective and analytic work.

Similarly, community engagement as an antifoundational practice forces students to question their certainties and as such expand their sense of the possible. For community-based experiences are (if we open ourselves up to it) truly destabilizing pedagogies that implode our grand narratives and fixed truths exactly because of their contingent character. And if community engagement can do that—if it can disrupt our sense of the normal to the extent that we internalize a "state of doubt"—then it exhibits Fish's other deeply desirous trait of higher education: the always restless and never closed-off search for truth.

These goals may at first seem modest, especially when compared to the often-heard claims that community engagement will transform higher education, the teaching and learning process, and local and global communities. But this is exactly Fish's point. The goals must be modest because the job is so complex. Education is an opening into the unknown; and careful, deliberate, and powerful education is extremely difficult to do well. Community engagement—a real-world and real-time pedagogy of engagement that confounds any simple or simplistic textbook notion of a fixed and stable truth—thus becomes a paradigmatic example of what Fish envisions as the ideal of higher education that he is trying to save.

This is actually an important realization, because it fully dovetails with Fish's argument that the academy must ultimately serve as a space for constant rethinking and defamiliarization; so long as this is done within the context of the academic (rather than political), any and all models for such inquiry are fair game. Which is why Fish (1994) approvingly cites feminism as a powerful intellectual force in and for higher education: "[T]he questions raised by feminism, because they were questions raised not in the academy but in the larger world and that then made their way into the academy, have energized more thought and social action than any other 'ism' in the past twenty or thirty years . . . [and] marks the true power of a form of inquiry or thought: when the assumptions encoded in the vocabulary of a form of thought become inescapable in the larger society" (294).

What Fish is approving is not a feminism in the academy that attempts to foster gender awareness or equity. This is not about cultural or political goals; rather, Fish sees feminist inquiry as strengthening higher education. Namely, Fish is always in search of the next "text" (broadly defined) that does "not allow a reader the security of his normal patterns of thought and belief" and a job well done in teaching it.

Community engagement—whether manifested as service-learning, public scholarship, community-based research, etc.—can and does serve such a function; it is a wonderfully complex and situated practice that truly disturbs and forces students (and faculty) to rethink their normal patterns of thought and belief. It brings to the fore the voices and practices of the community; it forces us to reconsider the very nature of scholarship, its practices, and its outcomes; it allows us to reimagine collaborative practices and interdisciplinary inquiry. And if this is so, then what we as practitioners and scholars must begin to do is work through how to teach it well.

The problem, though, as I have attempted to document throughout this book, is that all too often the rhetoric outpaces the reality. The problem—as many of the previous chapters have explored—is that community engagement is envisioned and positioned as a free-floating philosophy with wide-ranging institutional and pedagogical implications. I have shown this to be the case with service-learning and community engagement from both a theoretical perspective and pragmatically in regard to the lack of structures and academic guidelines and guideposts within many programs committed to service-learning practices. I have shown how recasting a field as a "public" practice—at least in the case of the field of sociology—has, for many scholars, undermined the very grounding and legitimacy upon which such a public stance was taken. Likewise, through the countermove of describing the formation and substance of interdisciplinary programs, I have shown how it was possible to ground the ideals and ideas of a social movement firmly within an intellectual one.

I thus want to suggest that Stanley Fish is right, that faculty need to believe that embracing community engagement is not antithetical to who they are as teachers and researchers. That it is and should be simply a part of their jobs. This is in direct contradistinction to the current reality that community engagement is for too many faculty just an overlay, and a poor one at that, onto our current job as teachers and scholars. It seemingly distracts us from and "adds" minimal "value" to our daily faculty work. I thus suggest that we must rethink community engagement as inherent to our positions and practices as good scholars and teachers. In one respect Boyer (1990) began this conversation almost two decades ago and it has been greatly expanded and enunciated (e.g., Huber and Hutchings 2005; Hutchings and Shulman 1999). My goal is to focus such work directly unto the minutia of faculty work and objectives.

Community engagement is not, in this perspective, so much about reconnecting institutions of higher education to the real-world lives of the students within it and the communities surrounding it. It is about the goal—at once much more humble yet much more radical—of providing faculty with an additional set of tools by which to carefully and effectively do their jobs. Put otherwise, it is exactly about embracing Fish's critique of focusing one's aim (rather than just "aiming low") through community-based models of teaching, learning, and research. By rethinking community engagement as the particular micropractices of faculty's day-to-day concerns and proclivities is to transform it from a silver bullet to a daily routine. This may not be rhetorically appealing, but it is pragmatically empowering. I want to thus engage in such a repositioning to support faculty buying into the process of a scholarship of engagement by providing a theoretically grounded vision of how to reframe its value and viability.

Four Models for Community Engagement

What I want to emphasize is that community engagement is not singular, that it in fact offers a diverse range of curricular and instructional options by which to engage in the issue deemed most valuable by a particular faculty in a particular course. To misappropriate Derrick Bell's terminology on racial integration, this is about interest convergence. It is about finding diverse means by which to accommodate diverse foci. In doing so, this not only brushes aside most traditional barriers to community engagement, but actually makes such a scholarship of engagement integral to individual and departmental goals.

It is thus useful to take the four distinctive typologies that I developed about service-learning in the first part of this book—technical, cultural, political, and antifoundational—and apply it to the varied means and goals that we may have for community engagement. Such typologies may of course overlap and intermix; yet each also has its own distinctive characteristics as well as limits and possibilities. This is what faculty need to understand so that they know what they are embracing (or rejecting) and the implications thereof.

Community engagement from a *technical* perspective is focused on its instrumental effectiveness, be it teaching, learning, or research. An excellent way, for example, to teach about the impact of poverty on families is to work with actual families in actual poverty within the context of an academic course that uses multiple other texts, reflections, and assignments. Undergraduate research or experiential education is here one amongst multiple pedagogical strategies; it serves the function of better teaching for better learning. A *cultural* perspective of community engagement is focused on the meanings of the practice for the individuals and institutions involved in the teaching, learning, and research practices. The specific practices are seen here as a means to, for example, help students increase their tolerance and respect for diversity, for academic institutions to promote engaged citizenship, and for local communities to overcome usually long-standing town/gown divisions.

Community engagement from a *political* perspective is focused on the promotion and empowerment of the voices and practices of historically disempowered and nondominant groups in society. Teaching, learning, and research are here viewed as the embodiment and enactment of a social-justice worldview, where the personal and the political meet in a substantive praxis and where higher education is viewed as a central agent of change for an equitable society. Finally, community engagement from an *antifoundational* perspective is focused on what John Dewey termed a "forked road" situation, one that fosters a state of doubt as a prerequisite for thoughtful deliberation. A scholarship of engagement that is antifoundational, given its embodied and enacted experiential components, serves as an opening for questioning seemingly "natural" norms, behaviors, and assumptions.

It should be noted that these typologies do not presuppose some teleological "great chain of being." It is just as legitimate to incorporate a community engagement component to teach mathematical principles as it is to develop cultural competency in future teachers. Additionally, any practice can have aspects from each typology. The semester-long tutoring of underperforming high school students in math can help college students understand how youth make systematic conceptual errors (a technical perspective); they can gain insight into working with youth of a different socioeconomic status and/or ethnic background (a cultural perspective); they can explicitly link the tutoring with college preparation to support college admission of such underrepresented youth (a political perspective); and they can begin to question to what extent academic success impacts identity formation (an antifoundational perspective). It all depends on the particular instructor, the course, and the curricular goals.

Each vision of community engagement, moreover, has embedded within itself its own limits and possibilities. Table 7.1 provides such an overview. The possibilities are, for the most part, obvious. Faculty members are intuitively drawn to community engagement when they realize that traditional pedagogical and research models may be inadequate to help them with their specific goals. All this typology does is better categorize their focus and legitimate that it *is* possible to engage the community in this particular way. Such a realization—that there are multiple ways by which engagement may be done—is itself a major realization for faculty unsure of how it is "supposed" to be done.

Table 7.1 Overview of four models for community engagement

	Key terms and focus	Possibilities	Limits
Technical	content knowledge	Cognitive success through real-world linkage	Experiential components may overwhelm the content focus
Cultural	civic engagement and cultural competency	Expansion of understanding of self in the local and global community	Complexity of social and cultural realities may be undermined by a "charity" orientation
Political	social & political activism	Fostering of a more equitable and socially just environment for individuals and groups	Ideology may appear partisan, thereby undercutting course content, student buying into community engagement, and achievement of rhetorical goals
Antifoundational	cognitive dissonance	Expansion of epistemological possibilities through questioning ofseemingly solid foundations	Lack of formal or conclusive solutions may disenfranchise committed students

The Limits of Community Engagement

But even more important, I believe, is that such a typology makes visible the limits of each mode of engagement. Articulating such limits assures faculty that they are not being sold a utopian bill of goods; that the theoretical and practical boundaries have been examined and thought through; and that there are definitive boundaries that can now be probed, examined, prepared for, and accepted. To support community engagement without clarifying such conceptual and practical limits is a foolhardy attempt since faculty's buying into the engagement is not about immaculate perceptions so much as it is about reasoned and reasonable disclosures. I thus summarize my analysis from Chapter 3.

Instructors working from a technical perspective should be aware that the experiential component of community engagement will raise a host of issues seemingly irrelevant to the course content being taught or the research study undertaken. Students engaged in math tutoring for a mathematics course, for example, may begin to ask about the seemingly sexist behavior of another student or teacher or they may question why so many underperforming youth are nonwhite. Instructors must then make difficult decisions as to whether to delve into such territory that may be far from their expertise, to ignore or minimize such discussions, or to develop entirely new readings that cut into already planned curriculum. A technical perspective, by its very nature of focusing on the content, leaves itself open to being undermined by the culturally saturated experiences of community engagement.

Instructors working from a cultural perspective face an analogous dilemma. What is intended to be a process of engaging (and hopefully embracing) the diversity and plurality of local and global communities may in fact reinforce students' deficit perspectives of the other. Students volunteering in a homeless shelter may see their worst stereotypes reinforced by violent, sexist, or demeaning behavior. They may now have "data," however anecdotal, that supports their predetermined convictions. The "border crossing" of a cultural perspective invites, as well, an implicit superior position of privilege that subverts the multicultural aims of coming to view and attempts to understand another way of being as equal and just as legitimate as one's own.

Whereas the limits to technical and cultural perspectives come from the actual engagement, instructors committed to a political perspective may find their limits already embedded within their own practices. Namely, a political perspective faces the dilemma of sliding into an ideological dogmatism that discounts alternative explanations or political worldviews. It is this dilemma that is at the heart of contemporary academic freedom debates. Conservatives believe that college students are only being exposed to one side of the academic story; and students may come to see community engagement exactly as such a partisan occasion. Moreover, given its strongly melioristic perspective, the political typology may find itself on the losing end of the tension between its utopian ideals and the limited pragmatic outcomes of an all too short fourteen-week semester.

Finally, instructors working from an antifoundational perspective may find their limits expressed not by resistant students (which is often seen from a political perspective), but from exactly those students who are committed to making a difference. It is a difficult lesson that asking the right questions may be more productive than making a grab for the nearest answers. An antifoundational perspective, as such, consistently struggles with not undermining itself as it struggles to work without grounding foundations. This is much like the story (told by Clifford Geertz) of the boy who asks his father, upon being told that the world rests on the back of a turtle, what did that turtle rest on? Another turtle, comes the answer. And that turtle? Well, it is turtles all the way down. It is hard to have passion and activism when one's foundations are unmoored.

STRENGTHENING FACULTY BUYING INTO COMMUNITY ENGAGEMENT

A scholarship of engagement has immense potential in higher education. As Derek Bok (2004) laid out in *Our Underachieving Colleges*, undergraduate education is a hodgepodge of unproductive practices that needs to be fundamentally altered to become relevant. Yet today's faculty are not trained or prepared to, for example, link their courses to their local communities or disseminate their research outside an academic audience. This is exactly Fish's point: that what faculty do well are the standard academic practices of investigating, analyzing, producing, and disseminating knowledge.

That is exactly why they will buy into community engagement when it is positioned as an integral component of their repertoire as productive scholars. If I have been searching for new methods to teach microeconomics (or gender theory or language acquisition or something else)—and have tried PowerPoint and teaching through films and project-based learning and wiki-based textbook construction—service-learning or community-based research may be yet another useful tool in my pedagogical toolbox. Such a technical perspective allows me to determine which particular ideas at particular points in the semester are best taught through community engagement. And which are not. Likewise, if I find that my students continue to insist, no matter how many texts I give them to read or how many different ways I present the lecture, that gender is "obviously" binary, community engagement with a local Lesbian Gay Bisexual Transgender (LGBT) organization may be a necessary component of my pedagogical strategy. Such an antifoundational perspective allows me to tailor my classroom practices to my academic goals.

I must of course be sure that in all such collaborations there is value and respect for the organization and that I am not perpetuating the "community as laboratory" phenomenon. Doing such legitimate community engagement is extremely difficult. But so is setting up a study that adequately controls extenuating variables or constructing a logical argument that both accounts for and surpasses alternative possibilities. That is, simply put, our job as academics.

Such a reconceptualization thus neatly answers Fish's critique by demonstrating that community engagement simply serves our (particular) goals of being good academicians who search for and develop ever better tools to conduct research and disseminate knowledge. Economics faculty may employ community engagement in a drastically different way from women's studies faculty, who in turn may have fundamentally different goals and methods than music education faculty. Yet so long as each discipline follows its own internal protocols and academic criteria for legitimate teaching and scholarship, it allows a thousand types of community engagement to bloom.

I have found that explicating the limits and possibilities of each mode of community engagement in this way is a necessary and powerful means by which to de-escalate all too prevalent implicit and explicit tensions of faculty concerned about the need to add "yet another" task onto their already too full plates. My argument instead is that faculty can continue to pursue their passions of "the search for truth"—be it teaching math or understanding social justice—so long as they do it through formal protocols that are respectful of their community partners, that follow legitimate pedagogical practices, and that are aligned to their particular academic goals.

Faculty's buying into community engagement is thus predicated on realizing that community engagement is just another component in our academic life of examining, questioning, researching, and synthesizing the knowledge and practices that work best for our particular subject matter. Put otherwise, a scholarship of engagement is no longer a free-floating add-on; it is no longer a political bellwether; and it is no longer an amorphous reform. It is a part of who we are as faculty.

There is thus no longer any need to save the world on our own time. Rather, a scholarship of engagement allows faculty to embed community-based models of teaching, learning, and research into the very structure and thus practices of our jobs. Fish's point—that we should focus on doing our job well in the confines of our own "shop"—is similar to the one developed in Chapter 4; namely, that only within the confines of an academic discipline can a field of study come to analyze, critique, and expand its founding assumptions, modes of practice, and boundaries of knowledge production and dissemination. Embracing community engagement takes this model one step further by suggesting that what we do as legitimate and careful scholars is the academic doppelganger to a social movement's attempt to "save the world." This is in no way a negation or dismissal of the potential power and force of social movements. It is simply a way to continue to play the game of academics within a different framework of engagement.

TOWARD A THEORY OF ENGAGED SCHOLARSHIP

Specifically, this different framework of engagement is one that embraces rather than ignores the pedagogical and experiential possibilities of teaching as a practice always already compromised by and imbued with the community-based, experiential, and embodied nature of engaged practice. To put it formally and conceptually, the teacher can never remove herself from the process of teaching; this is particularly true when experiential and field-based pedagogy is presumed (wrongly, as I have shown) to be synonymous with and conflated with being "the teacher." But experience is not transparent. As such, given this impossibility, a scholarship of engagement must foster and work with and through the implications: the academic embracing, examining, and giving of voice to the "remainder" always already inherent within the teaching, learning, and research of such practices.

As I have noted in a previous chapter, this argument rests in the Derridean and antifoundational point that the medium of instruction (be it the teacher, the text, or the service-learning experience) can never truly and totally "erase" itself; that is, there is no transparency. Technically, this is referred to as our inability to ever overcome the "economy of the sign." All texts and all teachers operate in this way. One can never simply attempt to add "balance" to the course readings to foster neutrality; nor can one remove oneself by arguing one is outside of such operations either because one is truly neutral on a topic or (alternatively) deeply and radically committed to one specific perspective. Instead, the Derridean point is that we always work within this economy. We as instructors can only attempt to claim neutrality by the very positioning we have within the system. As Bingham (2008) specifies: "The economy of erasure supplies us with the very currency with which we critique the economy of erasure" (21). The attempt by, for example, public sociology to distance itself from "professional sociology" to make a public difference must rely on the tacit notion that it can only do what it does because of its grounding in the formal discourse of sociology. We never erase ourselves.

This realization positions a scholarship of engagement in a fundamentally distinctive and different way. Specifically, it calls for the acknowledgment and thus examination of this lack of transparency; it suggests the criticality of engaging and strengthening the voices to be found within this engagement; and it suggests the need to embrace rather than remain distant from such engagement. To operationalize these theoretical insights into community engagement, I am suggesting that it is necessary to construct a conscious, careful, and critical academic examination of the process and product of a scholarship of engagement; that community voices must be central to the functioning of such community engagement; and that without the actual sustained and consequential engagement, there is no scholarship.

First, if my attempted teachings about the community—be it from a technical, cultural, political, or antifoundational perspective—can never rise above nor hide their particular assumptions, it becomes necessary to examine such assumptions. This does not suggest an inward and self-regressive examination. Rather, this is the standard and central aspect of methodological rigor and clarity that all disciplinary fields and discourses engage in when they develop, promulgate, and make use of a set of hermeneutic practices. Anthropologists self-reflexively speak of and examine the "researcher as instrument" and the problematics of writing for (and over) the native (Clifford and Marcus 1986). Black studies and Jewish studies scholars consciously and consistently return to the focus of who they are studying, who is allowed to study this "other," and the implications of such practices (Beroe 2005).

Second, such examinations cannot occur without the voices of the community. This is not a simple declaration of respecting and giving equal voice to the community that is a standard component of the service-learning movement. Rather, as Lori Pompa has done with her "inside students," all participants have a stake in the outcome. This does not mean that the community members or incarcerated men have a more legitimate or valorized perspective because they speak from "within" the community being engaged with. Rather, just like the immigrant detainees told Dicklitch's students their stories, so must the community be able to have a seat at the table to speak.

Finally, community engagement is encompassing to the extent that if there is no engagement, there is no scholarship. Because community engagement practices become embedded in and integral to the particular task (be it a methodology for teaching a course or a guiding principle of a department), the practices take on meaning contextually. One thus cannot, for example, remove a service-learning component from a course (as one would with an "add-on") without compromising the very heart of the course. Thus in exact contradistinction to the notion of service-learning as able to fit any and all disciplines and courses, such community engagement is highly particularistic and dependent on the specifics of the philosophical and pedagogical strategy and practice.

The implication of this line of argument is that rather than "saving the world on our own time," good engaged scholars can work with the world on the institution's time. We can simply do our job, and do it well, by making it self-reflexively critical, by providing a space for community response, and by being as comprehensive as our particular disciplinary rules and norms specify. The "list" of what I do thus becomes almost meaningless in comparison to how I do it. Whether I am engaged in community-based research or experiential education or service-learning is nowhere near as important as whether I have followed methodological rigor, whether I have exposed such practices to critique and reflection, and whether I have engaged in such practices with due diligence. This, I believe, even Stanley Fish would buy into.

Chapter 8

Living With(in) The Future: Higher Education Trends and Implications for Service-Learning

Throughout this book I have argued that service-learning, rather than transforming higher education, will be transformed by it. I have argued that higher education operates by specific disciplinary rules with highly particularistic academic norms and barriers; and for service-learning to survive and flourish would require a rethinking of how we thought about and enacted community-based models of teaching, learning, and research. This "disciplining" of service-learning, I have argued, should be both structural within academic programs and cultural about the nature of scholarship.

In this concluding chapter I want to push this argument one step further. I suggest that current models of service-learning may become anachronistic in the near future given recent and convergent trends of demographics, market pressures, and the nature of faculty work. Unless we are able to rethink service-learning within the academy, I suggest that the traditional ideals and practices of service-learning may isolate the very reform that advocates have hoped will spread across higher education.

This is not a doomsday scenario for engaged and activist pedagogical models and practices. Service-learning is a highly flexible and adaptable practice that works across an immense variety of institutions, faculty, disciplines, and students. It can accommodate differing and divergent goals and it is manifest in manifold instructional and institutional strategies, as I have documented in previous chapters. Service-learning's ability to function and spread across the panoply of higher education is of course part of what makes it such an appealing pedagogical and philosophical model. Likewise, the permutations of service-learning—for example, public scholarship, undergraduate research, community-based research—make it likely that multiple strands and types of engaged scholarship will be a part of the academy for a long time to come.

But while the practice of service-learning may be extremely flexible and accommodating to the extreme diversity of teaching, learning, and research in higher education, our thinking about its larger role and function may be much less so. Specifically, dominant conceptualizations of service-learning continue to privilege what I have earlier (following the example of feminism and women's studies) called a "first wave" mentality that sees service-learning as *the* answer, irrespective of what the question is. This book has been an attempt to strengthen a "second wave" of questioning and critique. My goal has been to provide a flexible academic underpinning that accommodates the strengths of service-learning instead of containing and constraining it.

In this final chapter I thus want to lay out what I see as the clear-cut trends within higher education that signal the marginalization of service-learning as traditionally envisioned and enacted as an overarching paradigm for transformation. The trends I describe are neither new nor unexpected; scholars have long documented such developments and the interlinkages amongst them (e.g., Altbach 1980; Clark 1987; Finkelstein 1984; Gumport 2000; Kirp 2003; Schuster and Finkelstein 2006; Slaughter and Rhoades 2004). Yet the implications—the changing role and nature of postsecondary education; the "delivery mechanisms" of such education; and the "clientele" to whom it is delivered—have been only minimally discussed within the context of and hope for the future of service-learning in higher education. Such trends should do more than give us pause; they should incite a current generation of scholars and practitioners committed to community-based practices and scholarship to reevaluate how engaged scholarship should be practiced. For while the service-learning scholarship continues to advance, it is almost completely focused on itself as a methodology rather than a mode of inquiry.

It is from this perspective that I consciously highlight trends that appear as hampering the vision and practice of service-learning. Enough literature is available that highlights and promotes the possibilities (e.g., Democracy Collaborative 2007; Ellison and Eatman 2008). I, instead, want to position such discussions within the context and thus in sync with the coming transformation of higher education.

DEMOGRAPHIC SHIFTS IN HIGHER EDUCATION

The most visible shift of the last half-century in higher education has been who actually goes to college. In 1965, a little less than six million students were enrolled in postsecondary institutions. Two-thirds of them were male, 70 percent full-time, and almost three-quarters were between the ages of eighteen and twenty-four. In 2005, almost eighteen million students were enrolled in postsecondary institutions. Just two-fifths of them were male, 60 percent full-time, and 60 percent were between the ages of eighteen and twenty-four (USDOE 2007a). This trend—of ever larger proportions of the general population enrolling in higher educational institutions—continues unabated. U.S. Department of Education projections suggest that by 2016, more than twenty million students will be enrolled (USDOE 2007b).

What is striking, though, are not simply the sheer numbers. Rather, it is the demographic diversity and goal-oriented plurality of these students that in turn affects the types of institutions and programs that are expanding. As Martin Trow (2005) has argued, higher education since World War II has been buffeted by the rate of growth, the absolute size of such growth, and the proportion of the relative age groups enrolled in higher education institutions. This shift, in turn, modifies the goals and practices of higher education from elite to mass to universal access that focuses on supporting whole populations' ability to prepare for workforce and technological changes.

For example, currently 40 percent of all students are enrolled in two-year institutions and 46 percent are enrolled as part-time students. Almost half of all students in postsecondary institutions are married, one-third work full-time, and more than a quarter of all students are over thirty years of age. Even when one examines full-time, traditional-age undergraduates, one finds "careerist" orientations whether one looks at their beliefs about the relevance of college or about the enrollment shifts from liberal arts to professional and preprofessional majors (Pryor et al. 2007). The "massification" of higher education is thus deeply linked to the privatization of education, that is, education as a private rather than a public good (Labaree 1997; Kezar et al. 2005).

The import of these trends is that there is no longer a "traditional" student seeking a so-called traditional liberal arts education. In fact, the vision of higher education occurring on quad-centered campuses filled by youthful students completing their studies in four consecutive years while being taught by faculty experts in their fields is a mirage (Butin 2005a). Outside of the hundred or so institutions at the top of the news magazines' rating games, such an education has never existed and certainly does not exist today. This does not mean today's higher education is irrelevant or outdated. In many ways the answer is quite the opposite, for today's postsecondary institutions are oftentimes much better at meeting the needs of current students than they were thirty or even ten years ago. These students and their needs, though, are not what we generally assume.

While I explore the implications below, it is noteworthy to draw out just one major implication of these demographic trends. Namely, most students in postsecondary education have minimal time or inclination to "do good" through traditional venues such as community service and service-learning. Thus while data of high school and traditional-age undergraduates show increasing patterns of community service (e.g., CIRCLE 2003), the overall patterns in higher education are in fact the opposite. National data (USDOE 2000) show that more than 65 percent of all undergraduates did not do any community service in 2000; this percentage rises to over 70 percent in two-year colleges and to over 80 percent in nondegree granting institutions. The only institutions where community service is prevalent are private nonprofit four-year institutions.

The splintering of goals and pathways across higher education suggests that any so-called movement attempting to (re)capture a democratic spirit or public good may be less a broad plank across higher education and more of a sliver. This movement's clarion call for a rejuvenated public sphere may resonate within the brick-and-mortar halls of the 286 liberal arts colleges (as designated by the Carnegie Foundation's classification system); but these institutions constitute just 6.5 percent of all postsecondary institutions and enroll just 3 percent of all its students. Of course the voices for a scholarship of engagement are not mute everywhere else, but these demographic trends suggest that they are certainly muted.

Market Pressures

This demographic transformation and institutional isomorphism to "client" needs is linked to anddriven by the marketization and segmentation of higher education. Much recent scholarship has focused on the "corporatization" of the academy (e.g., Bousquet 2007; Kirp 2003; Washburn 2006). From distinguished faculty chairs sponsored by Enron to the corporate buying of research and scholarship to the McDonaldization of curriculum construction by online programs, the migration of business perspectives and business practices to the academy has become pervasive.

One aspect of this transformation that I want to highlight is the standardization of the teaching and learning process within higher education. As Musselin (2007) has noted, the industrialization of higher education has become notable as the academy moves from "craft production of ad-hoc products to the organised production of mass products through the three mechanisms [of] specialization of tasks, rationalisation, and normalization" (182). Musselin is here noting the decoupling of tasks that used to be performed by a single faculty member into a production line of specialists, some of whom write curricula, others who change such curricula into online or other reproducible forms, others who teach such curricula, and still others who grade the student outcomes within the courses taught. This is a highly efficient and rational process that tightly links goals, objectives, and outcomes to produce replicable and quantifiable results that can, as such, be modified and thus made ever more effective and efficient.

In one respect this is a sought-after and desirable goal. As Zemsky et al. (2005) argue, such standardization is a highly positive development as it forces colleges to acknowledge that there is accountability for what is taught, how it is taught, and to what effect. Zemsky et al. (2005), for example, argue that the "near-absence of both structure and coherence from the American collegiate curriculum" was ruefully exposed once

> the University of Phoenix, in its embrace of the market, produced some of the most structured as well as coherent curricula in higher education . . . Phoenix made clear that teaching was a product to be owned, invested in, and supervised by the institution itself. Courses required management as well as design and then evaluation on a continuing basis. Consistency, borne of acknowledged standards, constituted successful teaching. (126)

This division of labor—whereby the construction of a course requires multiple skill sets (content expertise, technological skills, pedagogical competence)—neatly mimics business practices of replication.

It also, though, exacerbates and strengthens a course's adherence to a standardized outcome in that it "cannot be changed or adapted in real time, but also because they have to respect technical and conceptual norms. Teaching materials are no longer the personal handiwork of a specific teacher but more generic products that can be used by different tutors" (Musselin 2007, 182).

This is the normalization of market pressures. Successful teaching, in this light, is not the result of individual and dedicated instructor/artisans engaging in the craft of deep teaching and learning based upon a guild notion of what constitutes success. Rather, successful teaching is the consistency of knowing that any instructor, in any locale, can teach a first-rate predeveloped lesson that can be understood by students and that the outcomes of such understanding can be objectively measured. The University of Phoenix and Walden University (the former was founded in Arizona and the latter in Florida, but both are primarily online institutions) and a host of proprietary as well as nonprofit institutions have banked their existence on such economies of scale and division of labor.

The consistency of such curricular and pedagogical models in turn necessitates a limited and highly prescribed realm for teaching and learning. The normalization of such practices consists in the statistical reality that the further afield one moves from prescribed teaching models, the greater the chances that such teaching will not be demonstrably and positively quantifiable. McDonalds does not take risks with how it makes its burgers. And University of Phoenix instructors do not take risks with how they teach business economics.

The implication of market pressures for the service-learning field is that engaged scholarship is, to put it bluntly, a liability. Service-learning, as I have argued throughout, is in many ways the antithesis to standardized and consistent pedagogical models. Service-learning, as an antifoundational strategy of destabilization, should in fact be considered inconsistent to the extent that it undermines long-standing and calcified perspectives and ways of being. This may be deep and profound teaching and learning. But it is far from the goals envisioned by institutions committed to a different version of "quality."

Service-learning and engaged scholarship are time-intensive, high-risk, and profoundly unquantifiable undertakings (at least in the short-term). Service-learning practices cannot be decoupled from each other; nor can they be standardized. In many ways, engaged scholarship is the quintessential artisan undertaking of expert faculty cobbling together successful practices from the happenstance of the specific students in a particular class in a certain locale with available resources. This is certainly a masterly craft, but it is far from an integral component of market-driven higher education.

FACULTY WORK IN THE NEW ACADEMY

The "massification" and industrialization of higher education have in turn transformed faculty work. The proliferation of competing and varied postsecondary institutions and platforms, the multiplying modes of instructional delivery, and the market pressures on revenue streams have all transformed who teaches, in what format, and toward what outcomes.

The most visible impact of these changes has been on the actual composition of the professoriate, as postsecondary institutions move from reliance on full-time tenured and tenure-track faculty to faculty hired on a contingency basis. In fact, Schuster and Finkelstein (2007) argue that "in the near-millennium history of the academic profession, there has never been a time in which change is occurring so rapidly . . . [A] fair reading of the *rate* of change through the centuries makes clear that those changes in the tradition-bound, organizationally conservative university have been gradual, very gradual. In recent years, however, the pace of change throughout postsecondary education has accelerated sharply. The modern research university bears only faint resemblance to its early Twentieth Century predecessor" (2).

The professionalization of the professoriate by the end of the nineteenth century and the critical American Association of University Professors (AAUP) formulation in 1940 of tenure as the creation of a permanent faculty position led to a massive growth of full-time faculty in the mid-twentieth century. For example, in 1970, almost 80 percent of the faculty were full-time (NCES 2004). Today, less than half are in this category. National trends demonstrate that the standard new full-time faculty hire is in a non-tenure-track position (USDE 2007). In fact, Schuster and Finkelstein (2006) point out that more than 50 percent of all full-time hires have been of this type since 1993. This development, they argue, "marks a seismic shift in the types of faculty appointments that are being made—from essentially an unknown phenomenon in 1969 to an absolute majority of full-time appointments at the onset of the twenty-first century" (195). These new hires are much more likely (compared to new tenure-track hires) to be younger, female, without a terminal degree, and remain at a lower professorial rank (Cataldi et al. 2005). The expansion of two-year institutions and the rise of proprietary institutions and online degree programs have as such dramatically, and seemingly for the long-term, shifted hiring practices to focus on and rely upon adjunct faculty.

The loss of tenure-track lines and full-time faculty, though, is but the more visible half of this transformation in higher education. As Schuster and Finkelstein (2006) suggest, "the phenomenon of part-time appointments (and the creation of a huge—and growing—cadre of teaching-only faculty) is old news. What is far less understood—either in terms of measuring the phenomenon itself or in beginning to comprehend the implications of it—is the restructuring of full-time academic work and careers" (232). Schuster and Finkelstein are here referring to the rise and expansion of full-time "term-track" (as opposed to tenure-track) faculty differentiated by and specialized in teaching, research, or administrative functions. Such faculty are more likely to be differentiated into "teaching faculty" or "research faculty" rather than assumed and expected to undertake the standard tripartite task of teaching, research, and service.

These changes in faculty work arise in part from the market pressures discussed earlier in regard to proprietary and online institutions. The reliance on a division-of-labor business model does not include the notion of a "core faculty" who engages in service to the institution or enhances the long-term capacity of the institution and the academic field through research and scholarship. Such practices and services, if necessary at all, are within the purview of a "pay per view" model, whereby the institution pays specific experts to construct new courses, other experts to serve on admissions committees, etc.

Such a model is standard practice in the online education field that has at its core the grounding in economies of scale with the concomitant implications for faculty work. National data (Waits and Lewis 2003) demonstrate that online education is massively growing at for-profit and nonprofit institutions alike. In 2000, 56 percent of all Title IV degree-granting institutions were offering distance education courses that enrolled more than three million students, more than 80 percent of whom were at the undergraduate level; 19 percent of all higher education institutions offered certificates or degrees fully through online instruction (Waits and Lewis 2003). This data makes vivid that faculty work will continue to morph toward ever more specialized functions.

Especially relevant to and dovetailing with this changing nature of faculty work is the growing presence of educational policy discussions and demands for direct measures of student outcomes in higher education. The recent Spellings Commission (USDOE 2006) and the discussions and changes it has caused focus attention on so-called objective and cross-institutional measures of student learning. The Collegiate Learning Assessment (CLA), for example, is but one of the currently more popular standardized models that purport to examine and assess student growth over time through students' academic experiences (Klein et al. 2005, 2007). Through a ninety-minute test of its first- and fourth-year students, an institution can seemingly measure the value-added component of its educational practices.

The implication for faculty work is the continuing narrowing of functions, be it within a separate realm (of teaching, research, or service) or the specificity within each realm. "Teaching faculty" are not hired solely as expert pedagogues who can excite and focus the intellectual capacities of a roomful of undergraduates. Such faculty will be expected—much like the current "value-added" policy discussions at the K-12 level (Sanders and Horn 1998)—to demonstrate growth on standardized outcome measures that conform to predetermined benchmarks for seemingly objective criteria such as critical inquiry and logical thinking.

Such conformity of outcome measures may be a positive development for institutions that are attempting to codify and demonstrate their value to their paying clientele, but it bodes ill for ground-floor practices in the college classroom committed to dynamic pedagogical practices such as service-learning. Part-time and term-limited instructors hired to teach a particular course toward specific outcome measures—no matter how seemingly robust—are bound to be more risk-averse than tenured faculty teaching toward less quantifiable and narrow academic outcome measures. This is why, as I noted in a previous chapter, service-learning advocates have embraced "the quantitative move" to demonstrate the validity of their practice. Yet faculty work, especially among adjunct and specialized teaching faculty, is a delicate cost/benefit balancing act. Until and unless service-learning becomes predictable and efficient, it will not be embraced by those with minimal time or incentive for such an intensive undertaking.

And, to state but the obvious, if service-learning ever does become predictable and efficient, it will no longer be the transformational strategy envisioned and enacted by so many of today's advocates. This is truly the deep tension and difficult dilemma faced by service-learning and engaged scholarship today in higher education: how to maintain its vibrancy within the context of the changing face of higher education.

IMPLICATIONS FOR THE FUTURE OF ENGAGED SCHOLARSHIP IN HIGHER EDUCATION

Service-learning was forged within the fires of the civil rights movement. It was an attempt to break down the walls of the ivory tower to transform cloistered academic enclaves into vibrant centers of community revitalization. In many ways this linkage of academic practice with public relevance is at the forefront of how we want to think about higher education today; and in fact it may be convincingly argued that the vision and practices of service-learning are just as necessary and urgent today as they were forty years ago.

Whether the issue is neighborhood revitalization or a heightened understanding of and engagement with diverse communities, a scholarship of engagement presumes and enacts a collaborative and meaningful relationship between diverse constituencies. Service-learning, from this perspective, is but one of the more visible means by which faculty and students substantively connect with and attempt to improve their own and others' conditions. The linkage of academic knowledge and knowledge production with real-world relevance is crystal clear as soon as one opens the classroom or office door.

But such hopefulness, I suggest, cannot bear fruit amongst the roots being laid by the emergent trends described above. The changing face of student demographics, faculty profiles, and institutional structures and practices necessitates a fundamental rethinking of how a scholarship of engagement is to flourish in the academy. What does it mean for "nontraditional" students to engage in service-learning or community-based research? To what extent can or should "teaching faculty" be asked to embrace a time-intensive pedagogy that may not conform to predetermined curricula and not align with standardized outcome measures? Does the notion of "community service" or "public scholarship" even make sense within a for-profit college?

My goal in this book has not been to chart *the* future direction of the service-learning movement. Rather, my goal has been to make these kinds of questions visible and able to be asked to more clearly begin a fundamental rethinking of service-learning as an academic undertaking that truly belongs within higher education.

What higher education does, and does well, is the rigorous and critical analysis of issues worthy of study. If service-learning is to be a long-standing and productive component of this landscape, then it too must undergo such a process of rethinking. This is, yet again, an embrace of Stanley Fish's argument that we attend to the practices in our own shop and do them well. It is to suggest that one of the few things we in higher education can do is not save the world or do someone else's job, but serve as sites of analysis for critical and contested issues within contemporary culture. Service-learning as engaged scholarship does that very well. It is a cross-disciplinary, real-world, and consequential practice that fosters meaningful connections between colleges and their communities.

As I have argued throughout this book, though, such linkages are neither obvious nor predictable. Rather, engaged scholarship—for students, faculty, academic departments, or institutions writ large—must attend to its own practices, assumptions, and aspirations. If service-learning is to be sustainable and impactful, a second wave of practice, research, and critique must take hold.

Ultimately, service-learning and the scholarship of engagement may be extremely well-suited for the developing model and vision of the "engaged university" (Ellison and Eatman 2008) that is committed to public engagement and civic renewal. But to accomplish such goals, it is critical that we acknowledge and work within current higher education trends and the limits and possibilities inherent within contemporary models of service-learning. To not do so—to maintain the belief that goodwill and good work is enough—may ultimately undermine the service-learning movement much more than the trends of marketization, standardization, and job displacements described above. If we—as practitioners and scholars committed to engaged scholarship—believe in transforming higher education, then we must be willing as well to transform ourselves along the way.

References

Abes, E. S., G. Jackson, and S. R. Jones. 2002. Factors that motivate and deter faculty use of service-learning. *Michigan Journal of Community Service Learning* 9(1): 5–17.

Altbach, P. G. 1980. The crisis of the professiorate. *Annals of the American Academy of Political and Social Science* 448: 1–14.

American Association of Colleges and Universities (AAC&U). 1992. *Program review and educational quality in the major.* Washington, DC: AAC&U.

______. 2007. *College learning for the new global century.* Washington, DC: AAC&U.

American Association of University Professors (AAUP). 2007. *Freedom in the classroom.* http://www.aaup.org/AAUP/comm/rep/A/class.htm (accessed on August 28, 2008).

American Council of Trustees and Alumni. 2006. *How many Ward Churchills?* https://www.goacta.org/publications/downloads/ChurchillFinal.pdf (retrieved on November 16, 2009).

American Council on Education (ACE). 2005. *Statement on academic rights and responsibilities.* http://www.acenet.edu/AM/Template.cfm?Section=HENA&template=/CM/ContentDisplay.cfm&ContentID=10672 (accessed on June 23, 2005).

American Jewish Society Perspectives (AJS). 2006. *A forum on the Jewish studies undergraduate major: What do we learn about the field from how we educate our undergraduates?*, 8–31. http://www.ajsnet.org/ajsp06sp.pdf (accessed on August 21, 2008).

Angelo, T. A., and P. K. Cross. 1993. *Classroom assessment techniques: A handbook for college teachers.* 2nd ed. San Francisco, CA : Jossey-Bass.

Antonio, A. L., H. S. Astin, and C. Cross. 2000. Community service in higher education: A look at the faculty. *The Review of Higher Education* 23(4): 373–98.

Asante, M. K. 1998. *The afrocentric idea.* Rev. and expanded ed. Philadelphia: Temple University Press.

Astin, A. W., and L. Sax. 1998. How undergraduates are affected by service-participation. *Journal of College Student Development 39(3):* 251–63.

Astin, A. W., L. Sax, and J. Avalos. 1999. Long-term effects of volunteerism during the undergraduate years. *Review of Higher Education* 22(2): 187–202.

Baldwin, S. C., A. M. Buchanan, and M. E. Rudisill. 2007. What teacher candidates learned about diversity, social justice, and themselves from service-learning experiences. *Journal of Teacher Education 58*(4): 315–27.

Banks, J. A. 1996. *Multicultural education, transformative knowledge, and action: Historical and contemporary perspectives.* New York: Teachers College Press.

Barber, B. 1992. *An aristocracy for everyone.* Oxford: Oxford University Press.

Battistoni, R. 1997. Making a major commitment: Public and community service at Providence College. In *Successful service-learning programs: New models of excellence in higher education*, ed. E. Zlotkowski, 124–37. Bolton, MA: Anker Books.

Baxter Magolda, M. B. 1999. *Creating contexts for learning and self-authorship: Constructive-developmental pedagogy.* Nashville, TN: Vanderbilt University Press.

Becher, T., and P. Trowler. 2001. *Academic tribes and territories: Intellectual enquiry and the culture of disciplines.* 2nd ed. Philadelphia, PA: Open University Press.

Bell, C. A., B. R. Horn, and K. C. Roxas. 2007. We know it's service, but what are they learning? Preservice teachers' understandings of diversity. *Equity & Excellence in Education 40*(2): 123–33.

Bell, C., and H. Newby. 1971. *Community studies: An introduction to the sociology of the local community.* New York, NY: Praeger.

Bell, R., A. Furco, M. S. Ammon, P. Muller, and V. Sorgen. 2000. *Institutionalizing service-learning in higher education.* Berkeley, CA: University of California.

Bellah, R. N., R. Madsen, W. M. Sullivan, A. Swidler, and S. M. Tipton. 1986. *Habits of the heart.* Updated ed. Berkeley, CA: University of California Press.

Benson, L., I. Harkavy, and M. Hartley. 2005. Integrating a commitment to the public good into the institutional fabric. In *Higher education for the public good,* ed. A. Kezar, T. Chambers, and J. Burkhardt, 185–216. San Francisco: Jossey Bass.

Benson, L., I. Harkavy, and J. Puckett. 2007. *Dewey's dream: Universities and democracies in an age of education reform.* Philadelphia: Temple University Press.

Berle, David. 2006. Incremental integration: A successful service-learning strategy. *International Journal of Teaching and Learning in Higher Education* 18(1): 43–48. http://www.isetl.org/ijtlhe/pdf/IJTLHE65.pdf (accessed on August 23, 2008).

Biesta, G. 1998. The right to philosophy of education: From critique to deconstruction. In *Philosophy of education yearbook,* ed. S. Tozer, 476–84. Urbana-Champaign, IL: Philosophy of Education Society.

______. 2006. *Beyond learning: Democratic education for a human future.* Boulder: Paradigm.

Biesta, G., and N. C. Burbules. 2003. *Pragmatism and educational research.* Lanham, MD: Rowman and Littlefield.

Biglan, A. 1973. Relationship between subject matter characteristics and the structure and output of university departments. *Journal of Applied Psychology* 57: 204–13.

Bingham, C. W. 2008. Derrida on teaching: The economy of erasure. *Studies in Philosophy and Education* 27(1): 15–31.

Bligh, D. 2000. *What's the use of lectures?* San Francisco, CA: Jossey-Bass.

Bloorngarden, A. H., and K. O'Meara. 2007. Faculty role integration and community engagement: Harmony or cacophony? *Michigan Journal of Community Service Learning 13*(2): 5–18.

Bok, D. 2005. *Our underachieving colleges.* Princeton, NJ: Princeton University Press.

Borden, A. W. 2007. The impact of service-learning on ethnocentrism in an intercultural communication course. *Journal of Experiential Education 30*(2): 171–83.

Bousquet, M. 2008. *How the university works: Higher education and the low-wage nation.* New York, NY: New York University Press.

Bowles, G., and R. Klein. 1983. *Theories of women's studies.* London: Routledge and Kegan Paul.

Boyer, E. L. 1990. *Scholarship reconsidered: Priorities of the professoriate.* Stanford, CA: Carnegie Foundation for the Advancement of Teaching.

______. 1996. The scholarship of engagement. *Journal of Public Service & Outreach 1*(1): 11.

Boyle-Baise, M. 1999. "As Good As It Gets?" The impact of philosophical orientations on community-based service learning for multicultural education. *Educational Forum* 63(4): 310–21.

______. 2002. *Multicultural service learning: Educating teachers in diverse communities.* New York: Teachers College Press.

Boyle-Baise, M., B. Bridgwaters, L. Brinson, N. Hiestand, B. Johnson, and P. Wilson. 2007. Improving the human condition: Leadership for justice-oriented service-learning. *Equity & Excellence in Education 40*(2): 113–22.

Braxton, J. M., and L. L. Hargens. 1996. Variations among academic disciplines: Analytical frameworks and research. In *Higher education: Handbook of theory and research,* vol. XI, ed. John C. Smart, 1–46. New York: Agathon.

Breuning, M., and J. Ishiyama. 2004. International studies programs: For what purpose and for whom? A rejoinder to hey. *International Studies Perspectives* 5(4): 400–402.

Bringle, R., and J. Hatcher. 1995. A service learning curriculum for faculty. *The Michigan Journal of Community Service-Learning* 2: 112–22.

______. 2000. Institutionalization of service learning in higher education. *Journal of Higher Education 71*(3): 273–91.

Brown, J. N., S. Pegg, and J. W. Shivley. 2007. Consensus and divergence in international studies: survey evidence from 140 international studies curriculum programs. *International Studies Perspectives* 7(3): 267–86.

Brown, W. 1997. The impossibility of women's studies. *Differences: A Journal of Feminist Cultural Studies 9*(3): 79.

______. 2003. Women's studies unbound: Revolution, mourning, politics. *Parallax* 9(2): 3–17.

Brukardt, M. H., B. Holland, S. L. Percy, N. Simpher, on behalf of Wingspread Conference Participants. 2004. *Wingspread Statement. Calling the question: Is higher education ready to commit to community engagement?* Milwaukee: University of Wisconsin-Milwaukee.

Buechler, S. M. 1993. Beyond resource mobilization? Emerging trends in social movement theory. *Sociological Quarterly* 34(2): 217–35.

Burawoy, M. 2004a. 2004 American sociological association presidential address: For public sociology. *British Journal of Sociology* 56(2): 259–94.

______. 2004b. Public sociologies: Contradictions, dilemmas, and possibilities. *Social Forces* 82(4): 1603–18.

Butin, D. W. 2001. If this is resistance I would hate to see domination: Retrieving Foucault's notion of resistance within educational research. *Educational Studies* 32(2): 157–76.

______. 2002. This ain't talk therapy: Problematizing and extending anti-oppressive education. *Educational Researcher* 31(3): 14–16.

______. 2005a. Perspectives on higher education. *Educational Studies* 37(2): 157–66.

______. 2005b. Service-learning as postmodern pedagogy. In *Service-learning in higher education: Critical issues and directions,* ed. D. W. Butin, 89–104. New York, NY: Palgrave.

______. 2005c. Identity (re)construction and student resistance. In *Teaching social foundations of education: Contexts, theories, and issues,* ed. D. W. Butin, 109–26. Mahwah: NJ. Lawrence Erlbaum Associates.

______. 2006a. The limits of service-learning in higher education. *The Review of Higher Education* 29(4): 473–98.

______. 2006b. Disciplining service-learning: Institutionalization and the case for community studies. *International Journal of Teaching and Learning in Higher Education* 18(1): 57–64.

Calderón, J. Z. 2007. *Race, poverty, and social justice: Multidisciplinary perspectives through service learning.* Sterling, VA: Stylus.

Campus Compact. 2000. *Presidents' declaration on the civic responsibility of higher education.* Providence, RI: Campus Compact.

______. 2004. *2003 service statistics: Highlights of Campus Compact's annual membership survey.* Available at http://www.compact.org/wp-content/uploads/pdf/2004_statistics_summary.pdf (accessed on October 23, 2009).

Campus Compact, and Project on Integrating Service with Academic Study. 2000. *Introduction to service-learning toolkit: Readings and resources for faculty.* Providence, RI: Campus Compact.

Carnegie Foundation for the Advancement of Teaching. 2006. *Carnegie selects colleges and universities for new elective community engagement classification.* http://www.carnegiefoundation.org/news/sub.asp?key=51&subkey=2126 (accessed in December 2006).

Cataldi, F. E., M. Fahimi, and E. M. Bradburn. 2005. *2004 national study of postsecondary faculty (nsopf:04) report on faculty and instructional staff in fall 2003* (NCES 2005–172). U.S. Department of Education. Washington, DC: National Center for Education Statistics.

CIRCLE and Carnegie Corporation. 2003. *The civic mission of schools.* Washington, DC: CIRCLE.

Clark, B. R. 1987. *Academic life, small worlds, different worlds.* Princeton, NJ: Carnegie Foundation for the Advancement of Teaching; Lawrenceville NJ: Princeton University Press.

______, ed. 1987. *The academic profession: National, disciplinary, and institutional settings.* Los Angeles: University of California Press.

Clark, M. 2006. A case study in the acceptance of a new discipline. *Studies in Higher Education* 31(2): 133–48.

Clifford, J., and G. E. Marcus. 1986. *Writing culture: The poetics and politics of ethnography.* Berkeley, CA: University of California Press.

Cochran-Smith, M. 2001. Constructing outcomes in teacher education: Policy, practice and pitfalls. *Educational Policy Analysis Archives* 9(11). Available at http://epaa.asu.edu/epaa/v9n11.html (accessed on January 15, 2007).

Cohen, J. 2006. A laboratory for public scholarship and democracy. *New Directions for Teaching & Learning* 105: 7–15.

Colbeck, C. L., and P. Wharton-Michael. 2006. Individual and organizational influences on faculty members' engagement in public scholarship. *New Directions for Teaching & Learning* 105: 17–26.

______. 2006. The public scholarship: Reintegrating Boyer's four domains. *New Directions for Institutional Research* 129: 7–19.

Colby, A., E. Beaumont, T. Ehrlich, and J. Corngold. 2007. *Educating for democracy: Preparing undergraduates for responsible political engagement.* San Francisco, CA: Jossey-Bass.

Colby, A., T. Ehrlich, E. Beaumont, and J. Stephens. 2003. *Educating citizens: Preparing America's undergraduates for lives of moral and civic responsibility.* San Francisco: Jossey-Bass.

Coles, R. 1993. *A call to service.* Cambridge, MA: Harvard University Press.

Cooper, J. E. 2007. Strengthening the case for community-based learning in teacher education. *Journal of Teacher Education 58*(3): 245.

Cross, B. 2005. New racism, reformed teacher education, and the same ole' oppression. *Educational Studies 38*(3): 263–74.

Crow, G. 2000. Developing sociological arguments through community studies. *International Journal of Social Research Methodology: Theory and Practice* 3(3): 173–87.

______. 2002. Community studies: fifty years of theorization. In *Globalization and social capital*, special issue, *Sociological Research Online* 7(3). http://www.socresonline.org.uk/7/3/crow.html (accessed on August 21, 2008).

Cruz, N. I., and D. E. Giles, Jr. 2000. Where's the community in service-learning research? *The Michigan Journal of Community Service Learning*, issue 1: 28–34.

Cuban, L. 1990. Reforming again, again, and again. *Educational Researcher 19*(1): 3–13.

______. 1998. How schools change reforms: Redefining reform success and failure. *Teachers College Record 99*(3): 453.

Cuban, S., and J. B. Anderson. 2007. Where's the justice in service-learning? Institutionalizing service-learning from a social justice perspective at a Jesuit university. *Equity & Excellence in Education 40*(2): 144–55.

Deflem, M. 2007. Public sociology, hot dogs, apple pie, and Chevrolet. Inaugural issue, *The Journal of Professional and Public Sociology.* http://www.gsajournal.com/VolumeOneCover.pdf (accessed on August 28, 2008).

Democracy Collaborative. 2007. *Linking colleges to communities.* College Park, MD: Democracy Collaborative at the University of Maryland.

Derrida, J. 1976. *Of grammatology.* Baltimore, MD: The John Hopkins University Press.

Dewey, J. 1910. *How we think.* Boston: D.C. Heath and Co.

______. 1916. *Democracy and education.* New York: Macmillan.

______. 1938. *Experience and education.* New York: Macmillan.

Dicklitch, S. 2005. Human rights-human wrongs: Making political science real through service-learning. In *Service-learning in higher education: Critical issues and directions*, ed. D. W. Butin, 127–38. New York: Palgrave Macmillan.

Downing, D. B. 2004. Theorizing the discipline and the disciplining of theory. In *On anthologies: Politics and pedagogy*, ed. Jeffrey R. Di Leo, 342–72. Lincoln, NE: University of Nebraska Press.

Driscoll, A. 2008. Carnegie's community-engagement classification, intentions and insights. *Change* 40(1): 38–41.

DuBois, E. C. 1985. *Feminist scholarship: Kindling in the groves of academe.* Urbana: University of Illinois Press.

Ellison, J., and T. K. Eatman. 2008. *Scholarship in public: Knowledge creation and tenure policy in the engaged university.* Available at http://www.imaginingamerica.org/TTI/TTI_FINAL.pdf (accessed on July 23, 2008).

Eyler, J. 2000. What do we most need to know about the impact of service-learning on student learning? *The Michigan Journal of Community Service Learning* 7: 11–17.

Eyler, J., and D. Giles. 1999. *Where's the learning in service-learning?* 1st ed. San Francisco: Jossey-Bass.

Eyler, J., D. Giles, C. Stenson, and C. Gray. 2001. *At a glance: What we know about the effects of service-learning on college students, faculty, institutions and communities, 1993–2000.* Washington, DC: Learn and Serve America National Service Learning Clearinghouse.

Finkelstein, M. 1984. *The American academic profession: A synthesis of social science inquiry since World War II.* Columbus, OH: Ohio State University Press.

Fish, S. 1985. Consequences. *Critical Inquiry 11*(3): 433–58.

______. 1997. Boutique multiculturalism, or why liberals are incapable of thinking about hate speech. *Critical Inquiry 23*(2): 378.

______. 1999. *The trouble with principle.* Cambridge, MA: Harvard University Press.
______. 2003. Aim low. *Chronicle of Higher Education 49*(36): C5.
______. 2004. "Intellectual diversity": The Trojan horse of a dark design. *Chronicle of Higher Education 50*(23): B13–B14.
______. 2008. *Save the world on your own time.* Oxford, United Kingdom: Oxford University Press.
Flowers, K., and C. Temple. 2009. America reads as service-learning: A stereophonic report. In *Service-learning and the liberal arts,* ed. C. A. Rimmerman, 85–106. Lanham, MD: Lexington Books.
Foucault, M. 1977. Revolutionary action. In *Language, counter-memory, practice: Selected essays and interviews,* ed. D. Bouchard, 218–33. Ithaca, NY: Cornell University Press.
______. 1997. *Michel Foucault: Ethics, subjectivity, and truth.* Ed. Paul Rabinow. New York: The New Press.
______. 2000. Interview with Michel Foucault. In *Power,* ed. J. D. Faubion, 239–97. New York: New Press.
Freeland, R. 2004. The third way. *Atlantic Monthly* 294(3): 144–47.
Freire, P. 1994. *Pedagogy of the oppressed.* New York: Continuum.
Frickel, S., and N. Gross. 2005. A general theory of scientific/intellectual movements. *American Sociological Review* 70(2): 204–32.
Furco, A. 1996. *Expanding boundaries: Serving and learning.* Washington, DC: Corporation for National Service.
______. 2001. Advancing service-learning at research universities. *New Directions for Higher Education* 114: 67.
______. 2002. *Self-assessment rubric for the institutionalization of service-learning in higher education.* Berkeley, CA: University of California.
______. 2003. Issues of definition and program diversity in the study of service-learning. In *Studying service-learning: Innovations in educational research methodology,* ed. S. H. Billig and A. S. Waterman, 13–34. Mahwah, NJ: Lawrence Erlbaum Associates.
Furco, A., and S. Billig. 2002. *Service-learning: The essence of the pedagogy.* Greenwich, CT: Information Age Pub.
Gans, H. 1989. Sociology in America: The discipline and the public. American Sociological Association, 1988 presidential address. *American Sociological Review* 54(2): 1–16.
Gardner, H. 1983. *Frames of mind: The theory of multiple intelligences.* New York: Basic Books.
Geertz, C. 1973. *The interpretation of cultures.* New York: Basic Books.
Gelmon, S., B. Holland, S. Seifer, S. Shinnamon, and K. Connors. 1998. Community-university partnerships for mutual learning. *The Michigan Journal of Community Service-Learning* 5: 97–107.
Giroux, H. 1983. *Theory and resistance in education.* New York: Bergin and Harvey.
Goldhaber, D. D., and D. J. Brewer. 1999. A three-way error components analysis of educational productivity. *Education Economics 7*(3): 199.

Gray, M., E. Ondaatje, and L. Zakaras. 2000. *Combining service and learning in higher education.* Available at http://www.rand.org/pubs/monograph_reports/2009/MR998.pdf (accessed on September 18, 2009).

Gumport, P. 2000. Academic restructuring: Organizational change and institutional imperatives. *Higher Education* 39(1): 67–91.

______, ed. 2007. *Sociology of higher education: Contributions and their contexts.* Baltimore: The Johns Hopkins University Press.

Gumport, P., and S. Snydman. 2002. The formal organization of knowledge: An analysis of academic structure. *The Journal of Higher Education* 73(3): 375–408.

Guy-Sheftall, B., and S. Heath. 1995. *Women's studies. A retrospective: A report to the Ford Foundation.* Naugatuck, CT: Ford Foundation.

Hacking, I. 1999. *The social construction of what?* Cambridge, MA: Harvard University Press.

Hargreaves, A., L. Earl, and M. Schmidt. 2002. Perspectives on alternative assessment reform. *American Educational Research Journal* 39(1): 69–100.

Harkavy, I. 2006. The role of universities in advancing citizenship and social justice in the 21st century. *Education, Citizenship and Social Justice* 1(1): 5–37.

Hartley, M., I. Harkavy, and L. Benson. 2005. Putting down roots in the groves of academe: The challenges of institutionalizing service-learning. In *Service-learning in higher education: Critical issues and directions,* ed. D. W. Butin, 205–22. New York, NY: Palgrave Macmillan.

Harvey, I. E. 2000. Feminism, postmodernism, and service-learning. In *Beyond the tower: Concepts and models for service-learning in philosophy,* ed. C. David Lisman and Irene E. Harvey, 352–53. Washington, DC: American Association of Higher Education (AAHE).

Hatfield, G. 2002. Psychology, philosophy, and cognitive science: Reflections on the history and philosophy of experimental psychology. *Mind & Language* 17(3): 207–32.

Hayes, E., and S. Cuban. 1997. Border-pedagogy: A critical framework for service-learning. *The Michigan Journal of Community Service-Learning 4:* 72–80.

Head, B. W. 2007. Community engagement: Participation on whose terms? *Australian Journal of Political Science 42*(3): 441–54.

Henry, S. E. 2005. "I can never turn my back on that": Liminality and the impact of class on service-learning experiences. In *Service-learning in higher education: Critical issues and directions,* ed. D. W. Butin, 45–66. New York: Palgrave Macmillan.

Hey, J. A. K. 2004. Can international studies research be the basis for an undergraduate international studies curriculum? A response to Ishiyama and Breuning. *International Studies Perspectives* 5(4): 395–99.

Hillery, G., Jr. 1955. "Definitions of community: Areas of agreement." *Rural Sociology* 20: 111–22.

Himley, M. 2004. Facing (up to) "the stranger" in community service learning. *College Composition and Communication 55*(3): 416–38.

Hogan, K. 2002. Pitfalls of community-based learning: How power dynamics limit adolescents' trajectories of growth and participation. *Teachers College Record 104*(3): 586–98.

Holland, B. A. 2001. A comprehensive model for assessing service-learning and community-university partnerships. *New Directions for Higher Education* 114: 51.

hooks, b. 1994. *Teaching to transgress: Education as the practice of freedom.* New York: Routledge.

Horowitz, D. n. d. *Academic bill of rights.* www.studentsforacademicfreedom.org/abor.html (retrieved on November 7, 2006).

______. 2003. The campus blacklist. Available at www. studentsforacademic-freedom.org/essays/blacklist.html (accessed on November 3, 2009).

______. 2006. *The professors: The 101 most dangerous academics in America.* Washington, DC: Regnery Publishing.

Horowitz, D., and E. Lehrer. n. d. Political bias in the administrations and faculties of 32 elite colleges and universities. Available at http://www.studentsforacademicfreedom.org/reports/lackdiversity.html (accessed on August 17, 2008).

Huber, M. T., and P. Hutchings. 2005. *The advancement of learning: Building the teaching commons.* San Francisco, CA: Jossey-Bass.

Huberman, M. 1993. The model of the independent artisan in teachers' professional relations. In *Teachers' work: Individuals, colleagues, and contexts,* ed. J. W. Little and M. W. McLaughlin, 11–50. New York: Teachers College Press.

Hutchings, P., and L. E. Shulman. 1999. The scholarship of teaching: New elaborations, new developments. *Change 31*(5): 10–15.

Hyman, P. 2006. Forum response. *Association for Jewish Studies Perspectives* (Spring): 22–24.

Indiana University. 2008. Leadership, ethics, and social action. Available at http://www.indiana.edu/~lesa/ (accessed on March 3, 2008).

Ishiyama, J., and M. Breuning. 2004. A survey of international studies programs at liberal arts colleges and universities in the midwest: Characteristics and correlates. *International Studies Perspectives* 5(2): 134–46.

Jones, S. R. 2002. The underside of service learning. *About Campus 7*(4): 10.

Jones, S. R., J. Gilbride-Brown, and A. Gasorski. 2005. Getting inside the "underside" of service-learning: Student resistance and possibilities. In *Service-learning in higher education: Critical issues and directions,* ed. D. W. Butin, 3–24. New York: Palgrave Macmillan.

Kendall, J., ed. 1990. *Combining service and learning: A resource book for community and public service.* Raleigh, NC: National Society for Internships and Experiential Education.

Kezar, A. J., T. C. Chambers, and J. Burkhardt, eds. 2005. *Higher education for the public good: emerging voices from a national movement.* San Francisco, CA: Jossey-Bass.

Kirp, D. 2003. *Shakespeare, Einstein, and the bottom line: The marketing of higher education.* Cambridge, MA: Harvard University Press.

Klein, D. B., and C. Stern. 2005. How politically diverse are the social sciences and humanities? Survey evidence from six fields. In *Academic Questions*. Available at http://ssrn.com/abstract=664042 (accessed on November 12, 2009).

Klein, S. P., G. D. Kuh, M. Chun, L. Hamilton, and R. Shavelson. 2005. An approach to measuring cognitive outcomes across higher education institutions. *Research in Higher Education* 46(3): 251–76.

Klein, S. P., R. Shavelson, R. Benjamin, and R. Bolus. 2007. The collegiate learning assessment: facts and fantasies. Available at http://www.cae.org/content/pdf/CLA.Facts.n.Fantasies.pdf (accessed on October 29, 2009).

Kramer, M. 2000. *Make it last forever: The institutionalization of service learning in America*. Washington, DC: Corporation for National Service.

Kuhn, T. 1967. *The structure of scientific revolutions*. Chicago: University Of Chicago Press.

Kumashiro, K. K. 2002. Against repetition: Addressing resistance to anti-oppressive change in the practices of learning, teaching, supervising, and researching. *Harvard Educational Review 72*: 67–92.

Labaree, D. F. 1997. *How to succeed in school without really learning: The credentials race in American education*. New Haven, CT: Yale University Press.

______. 2004. *The trouble with ed schools*. New Haven, CT: Yale University Press.

Lather, P. 2005. Scientism and scientificity in the rage for accountability: A feminist deconstruction. Paper presented at the annual conference of the American Educational Research Association, Montreal, Canada.

Latour, B. 1979. *Laboratory life: The social construction of scientific facts*. Los Angeles: Sage.

Light, D. 1974. Introduction: The structure of the academic professions. *Sociology of Education* 47: 2–28.

Lindholm, J. A., K. Szelényi, S. Hurtado, and W. S. Korn. 2005. *The American college teacher: National norms for the 2004–2005 HERI faculty survey*. Los Angeles, CA: Higher Education Research Institute.

Lisman, C. D. 1998. *Toward a civil society: Civic literacy and service learning*. Westport, CT: Bergin and Garvey.

Liu, G. 1995. Knowledge, foundations, and discourse: Philosophical support for service-learning. *The Michigan Journal of Community Service-Learning* 2(1): 5–18.

Lounsbury, M., and M. Schneiberg. 2007. Social movements and institutional analysis. In *Handbook of institutional theory*, ed. R. Greenwood, C. Oliver, K. Sahlin, and R. Suddaby, 648–70. London: Sage.

Lucas, C. J. 1996. *Crisis in the academy: Rethinking higher education in America*. New York, NY: St. Martin's Press.

Lueddeke, G. 2003. Professionalising teaching practice in higher education: A study of disciplinary variation and "teaching-scholarship." *Studies in Higher Education* 28(2): 213–28.

Lyotard, F. 1984. The postmodern condition. Minneapolis, MN: Minnesota University Press.

Markus, G. B., and D. C. King. 1993. Integrating community service and classroom instruction enhances learning: Results from an experiment. *Educational Evaluation and Policy Analysis* 15(4): 410–19.

McKnight, J. 1989. Why "servanthood" is bad. *The Other Side* (January/February): 38–41.

Messer-Davidow, E. 2002. *Disciplining feminism: From social activism to academic discourse.* Durham, NC: Duke University Press.

Messer-Davidow, E., D. Shumway, and D. J. Sylvan, eds. 1993. *Knowledges: Historical and critical studies in disciplinarity.* Charlottesville, VA: University of Virginia Press.

Metzger, W. 1987. The academic profession in the United States. In *The academic profession,* ed. B. Clark, 123–208. Berkeley, CA: University of California Press.

Meyer, J., and B. Rowan. 1977. Institutional organizations: Formal structure as myth and ceremony. *American Journal of Sociology* 83(2): 340–63.

Michigan Journal of Community Service Learning (MJCSL). 2001. *Service-learning course design workbook.* Ann Arbor, MI: Office of Community Service Learning Press.

Mitchell, T. D. 2007. Critical service-learning as social justice education: A case study of the citizen scholars program. *Equity & Excellence in Education 40*(2): 101–12.

Moraga, C., and G. Anzaldúa. 1981. *This bridge called my back: Writings by radical women of color.* 1st ed. Watertown, MA: Persephone.

Morris, Aldon, and Carol M. Mueller, eds. 1993. *Frontiers in social movement theory.* New Haven, CT: Yale University Press.

Morton, K. 1995. The irony of service: Charity, project, and social change in service-learning. *The Michigan Journal of Community Service-Learning 2:* 19–32.

Mueller, C. M. 1992. Building social movement theory. In *Frontiers in social movement theory,* ed. Aldon Morris and Carol M. Mueller, 3–26. New Haven, CT: Yale University Press.

Mullins, N. C. 1973. *Theories and theory groups in contemporary American sociology.* New York: Harper and Row.

Musselin, C. 2007. Transformation of academic work: Facts and analysis. In *Key challenges to the academic profession,* ed. M. Kogan and U. Teichler, 175–90. Kassel, Germany: UNESCO Forum on Higher Education Research and Knowledge.

Naropa University. 2007. *Community studies center.* Available at http://www.naropa.edu/campuslife/csc/index.cfm (accessed on August 27, 2008).

National Center for Educational Statistics (NCES). 2000. *Youth service-learning and community service among* 6th–through 12th–grade students in the United States: 1996 and 1999. Washington, DC: Office of Educational Research and Improvement.

______. 2002. *Teaching undergraduates in U.S. postsecondary institutions: Fall 1998*. Washington, DC: U.S. Department of Education.

National Collaborative on Diversity in the Teaching Force (NCDTF). 2004. *Assessment of diversity in America's teaching force*. Available at http://www.ate1.org/pubs/uploads/diversityreport.pdf (accessed on November 2, 2009).

National Commission on Service-Learning (NCSL). 2002. *Learning in deed: The power of service learning for American schools*. Newton, MA: W. K. Kellogg Foundation.

National Council on Teacher Quality (NCTQ). 2004. *Increasing the odds: How good policies can yield better teachers*. Available at http://www.nctq.org/nctq/images/nctq_io.pdf (accessed on October 17, 2009).

National Survey of Student Engagement (NSSE). 2007. *Experiences that matter: Enhancing student learning and success*. Bloomington, IN: NSSE.

Neidorf, D. 2005. What's not served in service-learning. *Common Review* 4(2): 13–19.

Neumann, R., and T. Becher. 2002. Teaching and learning in their disciplinary contexts: A conceptual analysis. *Studies in Higher Education 27*(4): 405–17.

Northwestern University. 2008. *Certificate in service learning*. Available at http://www.sesp.northwestern.edu/ugrad/community/honors/certificate/ (accessed on March 3, 2008).

Nussbaum, M. 1999. The professor of parody. *New Republic* (February 22, 1999): 37–45.

O'Grady, C. R., ed. 2000. *Integrating service learning and multicultural education in colleges and universities*. Mahwah, NJ: Lawrence Erlbaum.

Paoletti, J. B., E. Segal, and C. Totino. 2007. Acts of diversity: Assessing the impact of service-learning. *New Directions for Teaching & Learning* 111: 47–54.

Paul, E. L. 2006. Community-based research as scientific and civic pedagogy. *Peer Review 8*(1): 12–15.

Peske, H. G., and K. Haycock. 2006. *Teaching inequality: How poor and minority students are shortchanged on teacher quality*. Washington, DC: Education Trust.

Pompa, L. 2002. Service-learning as crucible. *Michigan Journal of Community Service Learning* 9(1): 67–76.

______. 2005. Service-learning as crucible: Reflections on immersion, context, power, and transformation. In *Service-learning in higher education: Critical issues and directions*, ed. D. W. Butin, 173–92. New York: Palgrave Macmillan.

Portland State University. 2007. *Community studies cluster*. Available at http://www.pdx.edu/unst/sinq_communitystudies.html (accessed on August 27, 2008).

Powell , W., and P. J. DiMaggio, eds. 1991. *The new institutionalism in organizational analysis*. Chicago: University Of Chicago Press.

Prentice, M. 2007. Social justice through service learning: Community colleges as ground zero. *Equity & Excellence in Education* 40(3): 266–73.

Pryor, J. H., S. Hurtado, V. B. Saenz, J. L. Santos, and W. S. Korn. 2007. *The American freshman: Forty year trends.* Los Angeles, CA: Higher Education Research Institute.

Putnam, Robert. 2000. *Bowling alone: The collapse and revival of American community.* New York: Simon and Schuster.

Ramaley, J. A. 2006. Public scholarship: Making sense of an emerging synthesis. *New Directions for Teaching & Learning* 105: 85–97.

Rhoads, R., and J. Howard, eds. 1998. *Academic service learning: A pedagogy of action and reflection.* San Francisco: Jossey-Bass.

Rice, K., and S. Pollack. 2000. Developing a critical pedagogy of service-learning. In *Integrating service learning and multicultural education in colleges and universities,* ed. Carolyn O'Grady, 115–34. Mahwah, NJ: Lawrence Erlbaum.

Rich, A. C. 1979. *On lies, secrets, and silence: Selected prose, 1966–1978.* 1st ed. New York: Norton.

Richardson, V., and P. Placier. 2001. Teacher change. In *Handbook of research on teaching,* ed. V. Richardson and American Educational Research Association, 905–50. 4th ed. Washington, DC: American Educational Research Association.

Rojas, F. 2007. *From black power to black studies.* Baltimore, MD: Johns Hopkins University Press.

Rosenberg, P. M. 1997. Underground discourses: Exploring whiteness in teacher education. In *Off White: Readings on society, race, and culture,* ed. M. Fine et al., 79–89. New York: Routledge.

Rothman, S., S. Lichter, and N. Nevitte. 2005. Politics and professional advancement among college faculty. *The Forum* 3(1). Available at http://www.bepress.com/forum/vol3/iss1/art2 (accessed on September 25, 2007).

Rubin, D. 2005. Women's studies, neoliberalism, and the paradox of the "political." In *Women's studies for the future: Foundations, interrogations, politics,* ed. E. L. Kennedy and A. Beins, 245–61. Newark, NJ: Rutgers University Press.

Saltmarsh, J., and M. Hartley. 2008. *Framing Statement.* Invitational Colloquium, February 26–27, 2008. Dayton, OH: Kettering Foundation.

Sanders, W. L., and S. P. Horn. 1998. Research findings from the Tennessee value-added assessment system (TVAAS) database: Implications for educational evaluation and research. *Journal of Personnel Evaluation in Education* 12(3): 247–56.

Schnaubelt, T., and A. Statham. 2007. Faculty perceptions of service as a mode of scholarship. *Michigan Journal of Community Service Learning* 14(1): 18–31.

Schultz, B. D. 2007. "Not satisfied with stupid Band-aids": A portrait of a justice-oriented, democratic curriculum serving a disadvantaged neighborhood. *Equity & Excellence in Education* 40(2): 166–76.

Schuster, J., and M. Finkelstein. 2006. *The restructuring of academic work and careers: The American faculty.* Baltimore: Johns Hopkins University Press.

______. 2007. *On the brink: Assessing the status of the American faculty.* Berkeley, CA: Center for Studies in Higher Education.

Schutz, A. 2006. Home is a prison in the global city: The tragic failure of school-based community engagement strategies. *Review of Educational Research* 76(4): 691–743.

Shulman, L. 2004. *Teaching as community property: Essays on higher education.* San Francisco, CA: Jossey-Bass.

Sica, A. 2000. Rationalization and culture. In *The Cambridge Companion to Weber*, ed. S. P. Turner, 42–58. New York: Cambridge University Press.

Sigmon, R. 1979. Service-Learning. Three Principles. *ACTION* 8(1): 9–11.

______. 1994. Serving to learn, learning to serve. In *Council for Independent Colleges Report.* Washington, DC: The Council of Independent Colleges.

Slaughter, S., and G. Rhoades. 2004. *Academic capitalism and the new economy.* Baltimore: Johns Hopkins University Press.

Sleeter, C. E. 2001. Preparing teachers for culturally diverse schools: Research and the overwhelming presence of whiteness. *Journal of Teacher Education* 52(2): 94–106.

Sleeter, C. E., and C. A. Grant. 2003. *Making choices for multicultural education: Five approaches to race, class, and gender.* 4th ed. New York: J. Wiley and Sons.

Snyder, T. D., A. G. Tan, and C. M. Hoffman. 2004. *Digest of Education Statistics 2003.* U.S. Department of Education, National Center for Education Statistics. Washington, DC: Government Printing Office.

Sommer, J. W., ed. 1995. *The academy in crisis: The political economy of higher education.* Oakland, CA: Transaction Publishers.

Sperling, R. 2007. Service-learning as a method of teaching multiculturalism to white college students. *Journal of Latinos & Education* 6(4): 309–22.

Stacey, M. 1969. The myth of community studies. *British Journal of Sociology* 20(2): 134–47.

Stanton, D. C., and A. J. Stewart. 1995. *Feminisms in the academy.* Ann Arbor: University of Michigan Press.

Stanton, T., D. Giles, and N. I. Cruz. 1999. *Service-learning: A movement's pioneers reflect on its origins, practice, and future.* San Francisco: Jossey-Bass.

Stimpson, C. R., and N. K. Cobb. 1986. *Women's studies in the United States: A report to the Ford Foundation.* Naugatuck, CT: Ford Foundation.

Stoecker, R. 2002. Practices and challenges of community-based research. *Journal of Public Affairs* 6: 219–31.

Stoecker, R., Susan H. Ambler, Nick Cutforth, Patrick Donohue, Dan Dougherty, Sam Marullo, Kris S. Nelson, and Nancy B. Stutts. 2003. Community-based research networks: Development and lessons learned in an emerging field. *Michigan Journal of Community Service Learning* 9: 44–56.

Stoecker, R., and E. Tyrone. 2009. *The unheard voices: Community organizations and service-learning.* Philadelphia, PA: Temple University Press.

Swaminathan, A., and J. B. Wade. 2001. Social movement theory and the evolution of new organizational forms. In *The entrepreneurship dynamic in industry evolution*, ed. C. B. Schoonhovern and E. Romanelli, 344–58. Stanford, CA: Stanford University Press.

Swaminathan, R. 2005. "Whose school is it anyway?" Student voices in an urban classroom. In *Service-learning in higher education: Critical issues and directions*, ed. D. W. Butin, 25–44. New York: Palgrave Macmillan.

______. 2007. Educating for the "real world": The hidden curriculum of community service-learning. *Equity & Excellence in Education 40*(2): 134–43.

Tatum, B. D. 1992. Talking about race, learning about racism: The application of racial identity development theory in the classroom. *Harvard Educational Review* 62(1): 1–24.

Tittle, C. R. 2004. The arrogance of public sociology. *Social Forces* 82(4): 1639–43.

Tonnies, F. 1887/1954. *Community and society*. East Lansing: Michigan State University Press.

Trow, M. A. 2005. *Reflections on the transition from elite to mass to universal access: Forms and phases of higher education in modern societies since WWII*. Institute of Governmental Studies. Paper WP2005-4.

University of Kansas. 2008. *The process of certification in service learning*. Available at http://www.servicelearning.ku.edu/certification.shtml (accessed on March 3, 2008).

University of North Carolina-Chapel Hill (UNC). 2008. *Service scholars*. Available at http://www.unc.edu/cps/students-scholars-index.php (accessed on March 3, 2008).

U.S. Census Bureau. 2008. *U.S. Census Bureau News*. Available at http://www.census.gov/Press-Release/www/releases/archives/population/012496.html (accessed on October 17, 2009).

U.S. Department of Education (USDOE). 2004. *Meeting the highly qualified teachers challenge: The secretary's third annual report on teacher quality*. Washington, DC: USDOE.

______. 2006. *A test of leadership: Charting the future of U.S. higher education*. Washington, DC: USDOE.

Van de Ven, A. H. 2007. *Engaged scholarship: A guide for organizational and social research*. Oxford, United Kingdom: Oxford University Press.

Varlotta, L. E. 1997a. Confronting consensus: Investigating the philosophies that have informed service-learning's communities. *Educational Theory* 47(4): 453–76.

______. 1997b. A critique of service-learning's definitions, continuums, and paradigms: A move towards a discourse-praxis community. *Educational Foundations* 11(3): 53–85.

Vasta, E. 2000. *Citizenship, community, and democracy*. New York: St. Martin's Press.

Vaughn, R. L., and S. D. Seifer. 2004/2008. Recognizing service-learning in higher education through minors and certificates. In *Community-Campus*

Partnerships for Health. Available at http://www.servicelearning.org/instant_info/fact_sheets/he_facts/minors__certs_he (accessed on October 2, 2009).

Wade, R. C. 2007. Service-learning for social justice in the elementary classroom: Can we get there from here? *Equity and Excellence in Education 40*(2): 156–65.

Wahlke, J. C. 1991. Liberal learning and the political science major: A report to the profession. *PS: Political Science and Politics* 24(1): 48–60.

Waits, T., and L. Lewis. 2003. *Distance education at degree-granting postsecondary institutions: 2000–2001*. Washington, DC: U.S. Department of Education, National Center for Education Statistics.

Washburn, J. 2005. *University, Inc.: The corporate corruption of American higher education*. New York : Basic Books.

Weber, M. 1948. *From Max Weber: Essays in sociology*. Ed. H. H. Gerth and C. W. Mills. New York, NY: Routledge.

West, C. 1994. *Race Matters*. New York, NY: Vintage.

Westheimer, J., and J. Kahne. 2004. What kind of citizen? The politics of educating for democracy. *American Educational Research Journal 41*(2): 237–69.

Wharton-Michael, P., E. M. Janke, R. Karim, A. K. Syvertsen, and L. D. Wray. 2006. An explication of public scholarship objectives. *New Directions for Teaching and Learning* 105: 63–72.

Wiegman, R. 1999. Feminism, institutionalism, and the idiom of failure. *Differences: A Journal of Feminist Cultural Studies 11*(3): 107–26.

______. 2002. Academic feminism against itself. *NWSA Journal 14*(2): 1–7.

______. 2005. The possibility of women's studies. In *Women's studies for the future*, ed. E. L. Kennedy and A. Beins, 40–60. New Brunswick, NJ: Rutgers University Press.

Wilson, R. 2008. The public view of politics in the classroom. *Chronicle of Higher Education* 54 (30): A22. April 4.

Woodrow Wilson Foundation. 2005. *The responsive PhD*. Available at http://www.woodrow.org/images/pdf/resphd/ResponsivePhD_overview.pdf (accessed on March 28, 2008).

Young, C. A., R. S. Shinnar, R. L. Ackerman, C. P. Carruthers, and D. A. Young. 2007. Implementing and sustaining service-learning at the institutional level. *Journal of Experiential Education 29*(3): 344–65.

Zemsky, R., G. R. Wegner, and W. F. Massy. 2005. *Remaking the American university: Market-smart and mission-centered*. New Brunswick, NJ: Rutgers University Press.

Zlotkowski, E. 1995. Does service-learning have a future? *The Michigan Journal of Community Service-Learning* 2: 123–33.

Index

CPSIA information can be obtained at www.ICGtesting.com
Printed in the USA
LVOW10s0701241213

366413LV00003B/8/P

"Dan Butin is the kind of critical friend every service-learning advocate should excitedly embrace. This book is intellectually honest, theoretically sophisticated, and deeply impassioned. Butin embodies the kind of 'scholarly criticality' that anyone interested in moving the field forward should cherish. This book will shake your assumptions about service-learning in all the right ways."

—Joel Westheimer, University Research Chair and Professor, Faculty of Education, University of Ottawa, Canada

"This book is a deeply critical, reflective forum for considering questions related to where service-learning has come from and where it is going or ought to go. As such, it provides practitioners with an invaluable asset with which to inject a deeper criticality to theoretical and strategic debates and relate more collegially with the rest of the academy. I have felt for some time that to promote such reflection, service-learning needs a shake-up. Butin's path-breaking book is just what the doctor ordered."

—Tim Stanton, director of the Public Service Medical Scholars program, Stanford University School of Medicine and co-author of *Service-Learning: A Movement's Pioneers Reflect on its Origins, Practice, and Future* and *Engaged Scholarship Toolkit for Research Universities and Their Faculties*

This book offers a comprehensive rethinking of the theory and practice of service-learning in higher education. Democratic and community engagement are vital aspects of linking colleges and communities, and this book critically engages the best practices and powerful alternative models in the academy. Drawing on key theoretical insights and empirical studies, Butin details the limits and possibilities of the future of community engagement in developing and sustaining the engaged campus.

DAN W. BUTIN is the founding dean of the school of education at Merrimack College. He is the editor and author of over fifty books, articles, and book chapters, including the books *Service-Learning in Higher Education*, *Teaching Social Foundations of Education*, and *Service-Learning and Social Justice Education*. Dr. Butin's research focuses on issues of educator preparation and policy and community engagement. He has been the assistant dean of the school of education at Cambridge College and a faculty member at Gettysburg College. Prior to working in higher education, Dr. Butin taught middle school, an adult GED program, and was the chief financial officer of Teacher For America. More about Dr. Butin's teaching and scholarship can be found at http://danbutin.org/.

Cover design by Macmillan Publishing Solutions

palgrave
macmillan